RUBBER AND RUBBER
PLANTING

RUBBER AND RUBBER PLANTING

BY

R. H. LOCK, Sc.D.

Inspector H.M. Board of Agriculture and Fisheries
Sometime Scholar and Fellow of Gonville and Caius College
Cambridge and Assistant Director of Botanic
Gardens, Ceylon

Cambridge:

at the University Press

1913

CAMBRIDGE
UNIVERSITY PRESS

University Printing House, Cambridge CB2 8BS, United Kingdom

Cambridge University Press is part of the University of Cambridge.

It furthers the University's mission by disseminating knowledge in the pursuit of
education, learning and research at the highest international levels of excellence.

www.cambridge.org
Information on this title: www.cambridge.org/9781316601600

© Cambridge University Press 1913

First published 1913
First paperback edition 2015

A catalogue record for this publication is available from the British Library

ISBN 978-1-316-60160-0 Paperback

PREFACE

DURING recent years interest in the Rubber-planting industry has extended far beyond that comparatively large section of the community, which is engaged in trades more or less directly connected with rubber. In fact nowadays this material enters so intimately into the daily life of almost everyone, that there will probably be few to whom the romance of rubber entirely fails to make an appeal. In endeavouring to make this book suitable for the needs of as wide a circle of readers as possible, it has been the aim of the author to combine an accurate account of the scientific side of rubber planting with a certain amount of practical information which may be of use to the prospective planter. The space available in a book of this kind only admits of treating the subject in the form of an introductory outline, but it is hoped that the information given will be found reliable as far as it goes.

The science and practice of rubber planting are alike in their infancy, and in the immediate future important developments are to be anticipated both in our scientific knowledge of the physiological processes underlying the formation of latex, and in the practical methods of exploiting this most valuable of raw materials. It is partly owing to our want of knowledge that it is still possible to compress into a comparatively small compass a summary of what is accurately known of both branches of the subject.

The chapters on the physiology of latex are largely the outcome of original observations by the writer, whilst those on planting, harvesting and factory work on the estate are based on a close personal acquaintance with the industry in Ceylon. The chapter on disease, on the other hand, so far as it relates to the fungus pests of rubber, is little more than a summary of the work of Mr T. Petch, whose book is indispensable to anyone specially interested in this branch of the subject. I am also indebted to Mr Petch for the loan of the illustration of canker on *Hevea*. Free use has also been made of Mr Herbert Wright's well-known book on Para Rubber, and the student who wishes to enter further into the statistics of rubber cultivation will find therein a large mass of useful information.

Sir Daniel Morris's *Cantor Lectures*, delivered in 1898, still contain the best and fullest account of the wild sources of rubber, and I have drawn freely upon them for the information given in Chapter II.

The chemistry of rubber has been dealt with only in the barest outline, but the book seemed incomplete without some reference to this side of the subject. It is hoped that the brief summary of the processes employed in the manufacture of rubber goods will be of some interest both to the planter and to the general public. Of both these branches an excellent account on a much fuller scale is readily accessible in Dr Schidrowitz's book on Rubber. Other sources of information are acknowledged in a separate list of references.

I am indebted to several friends for advice and criticism. Mr W. N. Tisdall very kindly read through the whole of Chapters V, VI and VII, and made a large number of valuable suggestions, practically all of which have been incorporated. Mr Tisdall also provided the estimate given on p. 127. Professor T. B. Wood was also kind enough to read part of the proofs.

The text illustrations of different rubber-producing species have been drawn for me by Mr L. Denton Sayers —mostly from living specimens—and I am much indebted to him for the trouble he has taken with them. I have also to thank Mr H. F. Macmillan and Mr C. Northway for the loan of valuable photographs, Mr Staines Manders for the loan of blocks from the Catalogue of the New York Rubber Exhibition, and Messrs F. Shaw and Co. for illustrations of machinery.

I should not forget to mention the name of Mr C. O. Macadam my collaborator in the preparation of the Ceylon Handbook for the Rubber Exhibitions of 1911 and 1912, since this handbook is in a sense the nucleus from which the present volume has been evolved.

Finally it is a pleasure to acknowledge that any literary merit which this volume may possess is largely due to the untiring help and criticism of my wife, who has also devoted great care to the preparation of the index.

R. H. L.

1 *August* 1913.

TABLE OF CONTENTS

CHAPTER I

THE HISTORY OF THE USE AND CULTIVATION
OF RUBBER

Early uses. Vulcanised rubber. The discovery of rubber in America. The trade in wild rubber. Sir Clements Markham proposes plantations. Collection of seeds by Collins, Wickham and Cross. *Hevea brasiliensis* in Ceylon. Early experiments. Rise of the plantation industry in the East. Area under plantations. Production from estates. Variations in the price of raw rubber. Capital involved . . . . . pages 1—15

CHAPTER II

THE BOTANICAL SOURCES OF RUBBER

List of species. American species. *Hevea brasiliensis.* Economic aspects of the Amazon rubber industry. *Manihot* species. *Castilloa. Hancornia* and other species. Guayule rubber. African rubbers. *Funtumia. Landolphia.* Other African species. Asiatic rubbers. *Ficus elastica.* Jelutong rubber . . . . . . 16—37

CHAPTER III

THE PHYSIOLOGY OF LATEX PRODUCTION

The structure and functions of the vegetative organs. The laticiferous system. Latex tubes. Latex vessels. Structure of the bark of *Hevea.* Minute anatomy. The effects of wounding the bark. Renewal of bark. The functions of latex. The composition of *Hevea* latex. Coagulation.

38—55

CHAPTER IV

THE PHYSIOLOGY OF LATEX (*continued*)— TAPPING EXPERIMENTS

Method of experiment. Wound response. Reasons for the increase of yield. Duration of yield. Relation of yield to volume of bark. Origin of latex. Seasonal variation. Effect of climatic conditions. Variation in yield of individual trees. Tapping intervals. Overtapping. Yield at different levels of the trunk. Effect of tapping on the composition of the latex. General remarks on yield. Resting periods. Effect of tapping on the tree. Summary . . . . 56—92

CHAPTER V

HEVEA. PLANTING OPERATIONS

Choice of situation and soil. Clearing. Nurseries. Seed selection. Draining, irrigation, roads, etc. Lining and spacing. Holing and planting. Rate of growth. Weeding. Intercrops. Cultivation and manuring. Green manuring. Shade and wind belts. Pruning. Thinning out. Labour. Other expenses. Tools . . . . . 93—127

CHAPTER VI

HARVESTING OPERATIONS

Incision methods of tapping. Methods in use. Excision methods. Systems of paring. Marking the tree. Tapping. Angle of cut. Direction of cut. Distance between successive cuts. Yields at different levels. Tapping tools. Plantation yields . . . . . 128—152

CHAPTER VII

FACTORY WORK ON THE ESTATE

The factory. Transport of latex. Coagulation. Washing. Drying. Crêping. Smoking. Smoke curing of latex. Blocking. Scrap. Packing. The best form of plantation rubber. Sales and markets. Quality. Defects and blemishes. Tackiness . . . . 153—175

CHAPTER VIII

THE PESTS AND DISEASES OF *HEVEA*

Plantation conditions. Epidemics. Wind. Animals. Insects. *Termes Gestroi.* Fungus diseases. Diseases of the roots. *Fomes semitostus.* Brown root disease. Diseases of the stem. Canker. Pink disease. Die-back. Burrs and nodules. General sanitation . . 176—196

CHAPTER IX

THE CULTIVATION OF SPECIES OTHER THAN *HEVEA BRASILIENSIS*

Castilloa. Manihot. Funtumia. Ficus elastica . . 197—209

CHAPTER X

THE CHEMISTRY OF INDIA-RUBBER

Difficulties of study. Composition of technically pure rubber. Physical properties. Destructive distillation. Synthesis. Chemical constitution. Vulcanisation . . . . 210—219

CHAPTER XI

THE MANUFACTURE OF RUBBER GOODS

Summary. Washing. Drying. Mixing. Preparation of sheet rubber. Calendering. Miscellaneous articles. Rubber solution. Vulcanisation. Hot curing. Dry process. Bath process. Cold curing. Reclaimed rubber. Vulcanite. The testing of rubber goods . . . . 220—237

INDEX . . . . 238—245

SHORT LIST OF REFERENCES

1. CLAYTON BEADLE and H. P. STEVENS. *Rubber.*

2. Catalogue of the Second International Rubber and Allied Trades Exhibition. London, 1911.

3. Catalogue of the Third International Rubber and Allied Trades Exhibition. New York, 1912.

4. C. CHRISTY. *The African Rubber Industry and Funtumia elastica.* 1911.

5. *Circulars and Agricultural Journal of the Royal Botanic Gardens,* Ceylon. 1898–1912.

6. J. COLLINS. On the Commercial kinds of Indiarubber. *Journal of Botany.* 1868.

7. —— *British Manufacturing Industries, Guttapercha and India-rubber.* 1876.

8. *Encyclopædia Britannica,* ninth edition, Article on Indiarubber.

9. Sir CLEMENTS MARKHAM. The Cultivation of Caoutchouc-yielding Plants in British India. *Journal of the Society of Arts.* 1876.

10. Sir DANIEL MORRIS. Sources of Commercial India-rubber, Cantor Lectures. *Journal of the Society of Arts.* 1898.

11. T. PETCH. *The Physiology and Diseases of Hevea brasiliensis.* 1911.

12. PHILIP SCHIDROWITZ. *Rubber.* 1911.

13. CARL OTTO WEBER. *The Chemistry of India Rubber.* 1902.

14. J. C. WILLIS. *Agriculture in the Tropics.* 1909.

15. HERBERT WRIGHT. *Hevea brasiliensis or Para Rubber,* fourth edition, 1912.

16. *The Tropical Agriculturist* to 1912.

LIST OF ILLUSTRATIONS

PLATE   *To face page*

I. Fruits and Seeds of Rubber Plants . . . 17
II. Tapping wild *Hevea* 20
III. Forest scene, showing preparation of Hard Para
    rubber 22
IV. Anatomy of *Hevea* 46
V. *Hevea* Rubber on Swampy Land 95
VI. *Hevea* Rubber and Tea 114
VII. *Hevea* Tree pricked with Serrated Knife . . 133
VIII. Vacuum Drying Machine 163
IX. Canker of *Hevea* bark 188
X. Nodules in *Hevea* bark 193

FIG.   PAGE
1. *Hevea brasiliensis* 18
2. *Manihot Glaziovii* 24
3. *Castilloa elastica* 27
4. *Funtumia elastica* 31
5. *Landolphia Kirkii* 32
6. *Ficus elastica* 34
7. Transverse section of *Hevea* bark . . . 46
8, 9, 10, 11. Anatomy of *Hevea* . . . . Plate IV
12. Tapping areas 58
13. Lining at right angles to a given base . . 106
14. Square and hexagonal systems of planting . . 109
15. Tapping system—full herring-bone . . . 138
16. ,, ,, half ,, ,, . . . ,,
17. ,, ,, half spiral . . . . ,,
18. ,, ,, full ,, ... ,,
19. ,, ,, Basal V . . . . . ,,
20. Method of marking a tree for tapping . . 141
21. Tapping tools—gouge and farrier's knife . . 144
22. Hand Washing Machine 161

CHAPTER I

THE HISTORY OF THE USE AND CULTIVATION OF RUBBER

Early Uses.

PROBABLY no raw material of vegetable origin has been put to such multifarious uses as indiarubber. No other vegetable product has risen with equal rapidity from a position of comparative insignificance to one of the highest commercial prominence.

Although the use of rubber by natives of the Western Hemisphere is historically chronicled upwards of 400 years ago, indiarubber was first used in England in the eighteenth century, and then only in the first instance for removing the marks of black lead pencils. The first patent for the employment of rubber for waterproofing purposes was not taken out until 1791. The further development of this use is closely associated with the name of Thomas Hancock, of the firm of Charles Macintosh and Co.; but the modern extensions of indiarubber manufacture only became possible after the discovery by Nelson Goodyear in 1839 of the process of combining rubber with sulphur, which is known as *vulcanisation*. Goodyear took out a further patent in

1851 for the manufacture of vulcanite by more complete combination of rubber with sulphur.

Vulcanised Rubber.

The method of vulcanisation, which ranks among the most important of all industrial discoveries, was first found out in America by Goodyear. In England, Hancock shortly afterwards arrived at the same method independently. The following facts will help the reader to realise the far-reaching importance of vulcanisation.

Raw rubber becomes soft and sticky when heated, and when cooled beyond a certain point it becomes stiff and almost horny in consistency. Vulcanised rubber retains its physical properties almost unaltered over a range of temperatures extending from the freezing point to the boiling point of water. After prolonged immersion in water, raw rubber absorbs as much as 25 per cent. of its own weight of moisture. On the other hand, "The water absorption of vulcanised rubber is extremely small —certainly not large enough to appreciably affect the insulation of a rubber cable after five years' continuous immersion" (Weber). From these facts the enormous increase in the durability and general usefulness of vulcanised rubber at once becomes apparent. Moreover, according to the proportion of sulphur which has entered into combination with the rubber, the physical properties of the finished product can be made to vary from those of the softest elastic up to those of the hardest vulcanite.

The Discovery of Rubber in America.

As the reader may perhaps have already anticipated, the first notice of the use of rubber comes to us in the history of the voyages of Columbus. Columbus found that the natives of Hayti possessed among other amusements a game of ball. "The balls were of the gum of a tree, and although large, were lighter and bounced better than the wind balls of Castile."

A fuller account was given by Juan de Torquemada in 1615. This writer describes a tree, called by the natives Ulequahuitl (*Castilloa elastica*), which was held in high estimation in Central America. The method of collection of the rubber, which flows out as a milky white substance when the tree is wounded, is described, and also its coagulation by setting in calabashes and subsequent boiling in water. Sometimes the latex was simply smeared over the bodies of the collectors and allowed to dry—a method still employed by some primitive tribes. The rubber so prepared was used for making balls, and for shoes for tumblers and jesters, whose antics it assisted ; and a medicinal oil was extracted from it. Even at this early date the Spaniards themselves employed the milk for waterproofing their cloaks.

The first accurate account of *Para rubber* is given by C. M. de la Condamine, who visited the Amazon country on an astronomical mission in 1735. He describes various uses of rubber by the Omaquas Indians, including that of making syringes or squirts. These instruments appear to have played an important part in social gatherings and even in religious festivals. From this use comes the Portuguese name Pao di Xirringa, the syringe tree. Hence also are derived the familiar terms *Seringa* for rubber and *Seringueiros* for the labourers employed in the collection of this material.

The Trade in Wild Rubber.

The recent development of the trade in wild rubber may be traced in the following table, which shows the history for nearly a century of the most important kind, namely Para rubber, the produce of *Hevea brasiliensis*. Prior to the development of the planting industry in the East, the export of Para rubber from Brazil represented about half the world's total supply of the raw material.

TABLE I

Exports of Para Rubber from Brazil.

Year	Tons	Year	Tons
1827	31	1870	6,601
1830	156	1880	8,679
1840	388	1890	15,354
1850	1,466	1901	28,161
1860	2,671	1910	38,200

Wild rubbers from Africa and Asia did not begin to come into the market in large quantities until after the

Brazilian trade was well established. For example the exports from the Congo State rose from 30 tons in 1887 to 2000 tons in 1897. At the same date similar amounts were being exported from Lagos and from the Gold Coast. A thousand tons of rubber were however exported from British India as early as 1873.

Sir Clements Markham proposes Plantations.

Herbert Wright has called attention to the fact that it was Hancock who in 1834 first suggested the possibility of cultivating the best kinds of rubber trees in the East and West Indies. The suggestion arose on account of the difficulties which Hancock and his colleagues experienced even at that date in procuring a sufficient supply of raw material. The actual birth of the rubber planting industry, however, dates only from the seventies, and is specially associated with the names of Sir Joseph Hooker, at that time Director of the Royal Gardens, Kew, of Sir Clements Markham who occupied an important position at the India Office, and with those of the collectors Collins, Cross and Wickham. The success of the introduction of cinchona to the East ten years earlier led Markham about 1870 to take up the question of the introduction of rubber to India. The first step was marked by the preparation in 1872 of a report by James Collins, who had already published an excellent account of the wild species of rubber in 1868.

The story of the winning of the rubber seeds from America is one full of romantic interest, and speaks volumes for the enterprise and determination of the collectors. The first seeds of *Hevea* to arrive at Kew were probably those brought by Collins from the Amazon in 1873. In 1875 Cross was shipwrecked whilst on the way home with a consignment of *Castilloa* plants and seeds. Nevertheless he managed to preserve his precious collection and bring it safe to land. He was sent out again to collect *Hevea* seed in 1877 and was again successful. Although only a few of Cross's *Hevea* seedlings were preserved, there must by this time be a considerable number of trees growing in Eastern plantations which are directly descended from the survivors of this consignment. Cross was also responsible for the introduction of Ceara rubber to Kew about the same date—a less difficult feat owing to the greater powers of resistance possessed by the seeds of *Manihot*.

By far the largest and most important supply of *Hevea* seeds to reach Kew was, however, that brought home by H. A. Wickham in 1876. Wickham, who was resident at the time in the rubber country of the Amazon, was commissioned to supply seeds to the Indian Government; but the Brazilian authorities were naturally opposed to the export of the seed, and it was a remarkable chance which threw the required opportunity in Wickham's way. An ocean-going steamer trading to the Amazon was there abandoned by her supercargoes without a return freight. Wickham boldly chartered the steamer on behalf of the Indian Government, and all hands were pressed into the service of collecting seeds. The cargo once aboard was passed by the port authorities as botanical specimens, and upwards of 70,000 seeds were thus safely transported to Kew. Less than 4 per cent. of these seeds, however, germinated. The writer has Mr Wickham's personal assurance that these seeds came from full-sized forest trees actually being worked for rubber, which grew at a considerable distance from the river on forest-covered plateaux some hundreds of feet above flood-level. The often repeated statement that the parents of the rubber plantations had their origin in swampy ground liable to floods, may therefore be taken to be entirely without foundation.

Although the Government of India paid all the expenses connected with the introduction of Wickham's seedlings, Ceylon was selected as the site of their chief tropical nursery. A special garden at Henaratgoda, in the low country near Colombo, was opened to receive them, and here were set out some 2000 plants which arrived in Ceylon in 1876 in 39 Wardian cases by the s.s. *Duke of Devonshire*.

In the same year smaller consignments of plants of *Hevea brasiliensis* were despatched from Kew to Burma, Java, Singapore, and the West Indies. In 1877 plants were sent to Mauritius and West Africa, and in 1878 to Fiji.

Plants of *Hevea Spruceana* were first sent to Ceylon in 1883, but they do not appear to have survived.

Castilloa and *Manihot Glaziovii*, Ceara rubber, were also distributed by Kew to the same colonies at about the same date.

Hevea brasiliensis in Ceylon.

The principal nursery for the trees, which were to form the origin of the planting industry, was however at Henaratgoda, in Ceylon. Here flowers first appeared upon the trees in 1881, and in the same year Dr Trimen, the Director of the Botanic Gardens, commenced experiments in tapping. The plantation was thinned out in 1882, and in 1883 260 seedling plants were raised, most of which were distributed in Ceylon. In 1884 there were over 1000 trees at Henaratgoda, but it was found necessary to thin the plantation again in 1885, and we read of 450 fine trees existing in 1887. In 1893 about 90,000 seeds were distributed to planters in Ceylon from the Henaratgoda trees, and in subsequent years similar numbers were available. Seeds were also distributed on a considerable scale by the Ceylon Botanic Department to Malaya and elsewhere, and it is curious to remark that in recent years large consignments of seed have been sent back from Ceylon and Singapore to America and the West Indies for planting purposes. At the present day about 40 of the original trees survive at Henaratgoda, the largest being upwards of ten feet in girth.

Early Experiments.

Some of the earliest experiments in tapping planted rubber trees were begun by Trimen at Henaratgoda in 1888. One of the largest of the original trees was tapped a few times in alternate years. Tapping consisted in making gashes in the bark with a hammer and chisel, in imitation of the methods employed in the Amazon country. The recorded yields were as follows:

TABLE II

Yields of Rubber from one tree at Henaratgoda.

Year	lb.	oz.
1888	1	$11\frac{3}{4}$
1890	2	10
1892	2	13
1894	3	3
1896	3	0

Despite the fact that his first estimate of the probable yield of a rubber plantation was a very low one according to modern ideas, and although it was not considered safe in those days to tap the trees at an earlier age than 10 or 12 years, Trimen foresaw that a very handsome profit could be obtained from rubber planting, and strongly advocated the cultivation of this product in Ceylon in his report for 1888. In 1890 the Ceylon Forest Department opened an experimental plantation which was increased to some extent in subsequent years, but on estates little planting took place during the decade immediately following.

Rise of the Plantation Industry in the East.

In Ceylon in the eighties, when the coffee plantations were practically exterminated, some attention was paid to the cultivation of Ceara rubber, but difficulties of tapping soon caused this product to be almost entirely neglected. From 1900 onwards further trials were made with this species and with *Castilloa,* but it was soon found that neither was so well suited as *Hevea* for the conditions generally prevailing in the planting districts of Ceylon. In fact, except in Africa, the fortunes of the rubber planting industry are almost entirely bound up with those of the last-named genus. Even in the Dutch East Indies, plantations of Assam indiarubber (*Ficus elastica*) are now being cut down to make way for *Hevea brasiliensis.* Our further remarks apply therefore mainly to *Hevea.*

In 1890 about 300 acres had been planted with rubber in Ceylon, and in 1900 about 1750 acres. Planting continued steadily until 1904, when the area was estimated at 11,000 acres, and then came the historic rush into rubber which characterised the years 1905—1907. In 1906 the first World's Rubber Exhibition was held at the Royal Botanic Gardens, Peradeniya, and by the end of the year 100,000 acres had been planted. The present area under rubber in Ceylon may be estimated at upwards of 250,000 acres.

In the Federated Malay States the development of the industry was even more rapid. In 1897 rubber estates covered only 350 acres in Malaya. By the end of 1906 the area of rubber plantations was practically equal to that in Ceylon. In 1912 Wright estimated this area to have increased to 420,000 acres. Whereas in Ceylon a material proportion of the rubber has been planted through existing tea estates, practically the whole area under rubber in Malaya has been cleared of virgin forest.

After Ceylon and Malaya, the next most important centre for rubber cultivation is the Dutch East Indies. It is estimated that 150,000 acres have recently been planted in Java and 70,000 in Sumatra. The latter is largely in the hands of English companies. The greatest part of this rubber is *Hevea*, but considerable areas of *Ficus*, *Castilloa* and *Manihot* also exist in Java.

Although India was the country originally proposed for the site of a great planting industry, the early consignments of seeds—the first dates back to 1873—did not meet with much success, and little planting took place before 1900. At the time of writing, however, 40,000 acres have probably been planted with *Hevea* in Southern India.

In West Africa at the present time plantations probably consist about equally of *Hevea* and of the native *Funtumia*. In Angola and in Central and East Africa, on the other hand, Ceara rubber is beginning to be widely cultivated. In America, too, the natural sources of rubber are being widely supplemented by plantations. In Mexico large areas are cultivated with *Castilloa*, and at the present time the Brazilian Government is making great efforts to encourage the establishment of *Hevea* plantations.

Wright gives the following estimate of the world's planted acreage in 1912:

TABLE III

Acreage under Rubber in different countries.

Country	Acres
Malaya	420,000
Ceylon	238,000
Dutch East Indies, Borneo and Pacific Islands	240,000
South India and Burma	42,000
German Colonies	45,000
Mexico, Brazil, Africa and W. Indies	100,000

This estimate is probably rather under than over the mark. An estimate by Van den Kerckhove gives 220,000 for Mexico, 80,000 for Brazil, and 100,000 for Africa. His total for the world is 1,131,000 acres in 1912.

Production from Estates.

A large proportion of the acreage described above has been planted since 1906, and it will be readily understood that the production of plantation rubber is rapidly increasing, and is likely to increase at a still more rapid rate in the immediate future. Indeed, at the present time the increase is in geometrical progression, and for some years past the output of rubber both from Ceylon and from the Federated Malay States has practically doubled every year. In fact, so far as Ceylon is concerned, the following table shows that this is an understatement of the case.

TABLE IV

Exports of Rubber from British Colonies in the East.

Year	Ceylon exports, tons	Malay Peninsula exports, tons
1906	147	425
1907	248	1,036
1908	407	1,665
1909	666	3,340
1910	1,472	6,500
1911	2,900	11,000
1912	6,300	about 20,000

Lewton Brain in 1910 estimated the future yields for Malaya as follows :

TABLE V

Estimated exports of Rubber from Malaya.

Year	Tons
1911	10,950
1912	18,750
1913	26,550
1914	35,640
1916	65,000

There seems every reason for anticipating that these estimates will be exceeded.

A million acres of rubber in full bearing may be expected at a rough estimate to produce 150,000 tons of rubber annually. If the present increase in the rate of consumption is maintained, it seems probable that the whole of this amount will be required as soon as the plantations can produce it. When this takes place, a considerable reduction in the consumption of the lower grades of wild rubber may be expected. Some reduction in the very large amount of reclaimed rubber used by manufacturers may also occur when high grades of fresh rubber are obtainable at a somewhat lower price. Hitherto the increased demand has been such that with the supplies of wild rubber practically stationary, or even somewhat increasing, all the plantation rubber available has been readily taken up, and prices have been maintained at what may still be regarded as a more or less artificial figure.

Variations in the Price of Rubber.

The history of prices is indeed remarkable, and the factors which rule them are not always easy to understand. Taking the price of fine hard Para as the standard, the value of this commodity in 1871 was about three shillings a pound. In 1878 the price was only two shillings. This rose to 4/4 in 1883, and fell to 2/6 in 1885. From this point the rise in value was more or less continuous until the price reached 5/8 in 1905. In 1908, a year of considerable industrial depression, the price fell to 2/9, but after this it rose very rapidly until it reached the unprecedented figure of 12/9 in

April, 1910. Much of this increase was probably associated with the increasing demand for motor tyres. After this, prices declined rapidly, and during 1912 the price remained fairly steady, between 4/– and 4/6.

The price of the best grades of plantation rubber is generally very close to that of hard Para. Hitherto, however, Para has always retained a higher value than the best plantation rubber, taking account of the fact that the former contains upwards of 10 per cent. of water more than the latter.

The amount of capital embarked in the rubber planting industry is enormous. Wright estimates that the nominal capital of companies registered in Great Britain alone, between 1907 and 1911, exceeds £90,000,000, of which about £60,000,000 is actually paid up.

CHAPTER II

WE propose to give in the following chapter some description of the different species of plants from which rubber is obtained, and of the methods employed in collecting the rubber from these plants in the wild condition. It will be convenient to consider the various species more or less in the order of their geographical distribution. Before doing so we may briefly enumerate the principal plants concerned according to their botanical classification.

List of the Principal Rubber-yielding Species.

I.   MORACEAE.

Castilloa elastica, C. Markhamiana.

Ficus elastica, F. Vogelii and other species.

II.   EUPHORBIACEAE.

Hevea brasiliensis and other species.

Manihot Glaziovii, M. dichotoma, M. heptaphylla, M. piauhyensis and other species.

Sapium Jenmani and other species.

Plate I

Photo H. F. Macmillan

Fruits and Seeds of Rubber Plants

III. APOCYNACEAE.

Alstonia sp.
Carpodinus sp.
Clitandra sp.
Funtumia elastica, F. africana.
Dyera.
Hancornia speciosa.
Landolphia Kirkii, L. Dawei, L. owariensis,
L. Heudelotii, L. Thollonii, etc.
Leuconotis sp.
Mascarhenasia elastica.
Parameria glandulifera.
Urceola esculenta.
Willughbeia sp.

IV. ASCLEPIADACEAE.

Cryptostegia grandiflora.
Raphionacme utilis.

V. COMPOSITAE.

Parthenium argentatum.

AMERICAN SPECIES.

Hevea brasiliensis, Para Rubber.

About 21 different species of *Hevea* are found in the Amazon region of North-West Brazil. Of these *Hevea brasiliensis* is the most important, and furnishes the highest quality of rubber.

Hevea brasiliensis is a tall and handsome tree, often reaching a height of 90 feet, whilst the circumference near the ground may exceed 12 feet. Commercially, its most important feature is the bark, which reaches a thickness of an inch or more in well-developed trees.

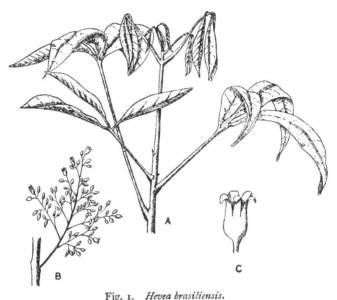

Fig. 1.   *Hevea brasiliensis.*
A.   Young shoot.   B.   Inflorescence.   C.   Flower.

In addition to containing an abundant store of latex, yielding from 30 to 40 per cent. of rubber, the bark of *Hevea* possesses an admirable consistency for the passage of the various tools used in tapping, combined with a remarkable faculty for recovering from the effect of wounds. The leaves are smooth, with three spear-shaped leaflets, and are very variable in size; indeed, all the characters of the tree are subject to marked variation. In Brazil the shape of the leaves is considered to be a feature by which good and bad varieties can be distinguished. The variety introduced into the East appears to be one of the best, although showing in its turn considerable variation. In Brazil the leaves fall from the trees about June, at the beginning of the so-called dry season, and are replaced by new leaves shortly afterwards. For a few weeks or days, therefore, the trees winter with bare branches.

The trees introduced into Ceylon have encountered there a reversed arrangement of seasons. On the Western side of Ceylon, accordingly, leaf-fall takes place in February.

The flowers appear shortly after the new leaves. They are small and greenish in colour, and possess a pleasant scent comparable with that of the blossom of the lime tree. The male and female flowers are separate, but are grouped in the same panicles. The fruits ripen in about five or six months. Three large seeds, marbled with brown and grey, are contained in a single fruit—a hard woody capsule, which bursts open when ripe and scatters the seeds to a considerable distance.

The structure and physiology of the bark of *Hevea* will be dealt with in some detail in the following chapters. We may proceed to describe the collection of the rubber in the forests of Brazil.

The trees are said to flourish best on rich alluvial clay swamps by the side of rivers. Tapping is carried out between June and February, and is confined to trees of 10 to 15 years of age and upwards. To each seringueiro or collector are assigned from 100 to 150 trees, which are connected by a winding path, or *estrade*, cut through the undergrowth. Tapping is performed with a hatchet, shaped like a small poleaxe, and the latex is collected in small cups of tinplate, having a sharp edge which can be inserted into the bark. Some days before the tapping proper is begun, the trees are gashed high up with a long-handled hatchet, in order " to make the latex ascend from the roots." The actual tapping is done on alternate days. Beginning at sunrise the seringueiro makes two rounds of his estrade, the first in order to tap the trees and the second in order to collect the latex. The first day's tapping consists of a horizontal row of slanting incisions made with the hatchet as high up as the operator can conveniently reach. On subsequent days fresh rows of cuts are made below the old ones, the bottom of the trunk being reached on the average after about 35 tapping days.

The next operation is smoking, which is carried out as soon as the latex is brought in, before coagulation and putrefaction have time to assert themselves. A smoky fire of wood, mixed with the nuts of certain palms, is made beneath a kind of funnel-shaped chimney. The latex is rotated in the smoke on the surface of a wooden instrument which may be either rod- or paddle-

Plate II

Tapping Wild *Hevea*

shaped. As each successive layer of rubber dries, fresh latex is poured upon the surface until a large block or ball of rubber is formed, which may weigh from 50 to 100 pounds.

The method above described appears to be nearly identical with that which has been employed for centuries by the natives of the Amazon valley. The rubber obtained thereby is still adopted as the highest standard of quality in the world's markets.

Economic Aspects of the Amazon Rubber Industry.

The country which produces Para rubber lies in the States of Amazonas, Para and Acre. The pioneers of the industry are constantly pushing further afield up the different tributaries of the Amazon. When the country has been explored and the presence of a suitable number of *Hevea* trees ascertained, a grant of land is obtained from the Government of the State, and after the payment of certain taxes, the land becomes private property.

The owner of the property is usually not himself a large capitalist. In order to finance his enterprise, he obtains a loan at a high rate of interest from a trader or rubber merchant. Labour is chiefly imported from the barren region of Ceara. The labourers often arrive at the base of the expedition in a practically destitute condition, and provision has to be made in advance for food supplies, tools and transport. All supplies have to be obtained from traders—naturally at high prices— and all must be carried long distances by steamer to the scene of operations. For working a seringal, or estate, of 200 estrades, an advance of 180,000 milreis, or nearly £10,000, including interest, may be required.

The paths of the seringal are practically sublet to the individual seringueiros, who have to pay for their food, tools, transport, etc., at high prices, together with interest on the loan advanced to them for incidental expenses, out of the value of the rubber which they obtain. Since the rubber must be sold by the owner to the trader, and by the trader to the exporting firm, very little profit is generally left for the individual collector. The inland freights, moreover, are very high, and an export tax has to be paid on the rubber at the rate of nearly 20 per cent. *ad valorem.*

The successful competition of the plantation industry in other parts of the world has recently led to active legislation on the part of the Brazilian Government, with a view to removing as far as possible the handicaps under which the wild rubber industry has hitherto laboured. These fall mainly under the heads of expensive labour, heavy transport rates and high export duty. So long as the price of rubber remains high, the Brazilian capitalist is able to pay for the high cost of production plus the high freights and taxes, but as soon as the price of rubber falls below three shillings a pound, the pinch will be severely felt, not only by the individual owner but also by the country in general, for

Plate III

Forest scene, showing preparation of Hard Para Rubber

the latter depends largely upon the export of rubber for
its revenue.

Special concessions are now being made to encourage
the introduction of foreign capital and the opening up
of plantations, by the granting of premiums and re-
mission of taxes.   In addition, the Federal Government
is taking steps :

1.   To encourage rubber collection and cultivation
by advancing money on easy terms, and by the offer of
premiums on new plantations.

2.   To encourage industries for the manufacture of
rubber goods—also by the offer of premiums.

3.   To help the regular workers and immigrant
labourers by the provision of special boarding estab-
lishments and hospitals and by the sale of stores at cost
price.

4.   To reduce the cost of transport by opening
railways and improving the navigation of rivers.

5.   To encourage the production of foodstuffs in the
Amazon Valley, in order to obviate the necessity of
transporting such materials for long distances.

6.   To hold triennial exhibitions in Rio de Janeiro
of everything relating to the rubber industry of the
country.

And last but not least to open agricultural experiment stations in different parts of the country[1].

[1] Abridged from the Handbook of the Third International Rubber and
Allied Trades Exhibition, New York, 1912.

Manihot Glaziovii.

This species, the source of Ceara rubber, extends over a wide area in North-Eastern Brazil. It is adapted to a drier climate than *Hevea*, and flourishes on a dry and rocky soil up to an altitude of about 4000 feet. The

Fig. 2. *Manihot Glaziovii.*
A. Leaves. B. Inflorescence. C. Female flower.

plant is comparatively shrubby, branching close to the ground and rarely exceeding 30 feet in height. The bark has a hard surface, and readily peels off from the trunk like that of a cherry or birch tree.

Tapping is therefore a more difficult process than in the case of *Hevea*, although the latex and rubber obtained from the trunk are closely similar both in quality and—relatively to the size of the tree—in quantity. The raw rubber contains somewhat more resin than *Hevea* rubber, and, however carefully prepared, is quite distinguishable from the latter. The latex coagulates without acid, on the mere addition of water.

The leaves are generally three-lobed, and afford a very dense shade, but the trees are markedly deciduous and remain bare of leaves for a considerable period. The seeds, which resemble those of the castor oil plant, are very hard. They retain their vitality for a long period, and do not germinate readily unless the shell is filed through. The roots are tuberous like those of the allied species *Manihot utilissima*, the cassava plant.

Owing to the difficulties which attend the tapping of the stem, the wild Ceara rubber is often tapped at the roots or very close to the ground. The latex is then simply collected in small cavities in the ground, opened wherever the subterranean organs are discovered. The cavities are usually lined with clay or leaves, but such a method of collection naturally leads to the production of a very impure form of cake or lump rubber. "Ceara scrap" is obtained by gashing the stems with a knife. The latex oozes out and is allowed to dry upon the bark. It is then pulled off in strings, which are either rolled up into balls or put into bags in loose masses.

Other Species of Manihot.

In 1907 extensive reports were published in Germany emphasising the importance of *Maniçoba rubber* as distinguished from ordinary Ceara. This rubber was said to be derived mainly from three species of *Manihot*—*M. dichotoma*, *M. heptaphylla* and *M. piauhyensis*. Formerly it appears to have been confused with the produce of *Hancornia speciosa*. The seed of the above-named species of *Manihot* has been widely advertised for planting in tropical countries, but the small amount of evidence at present available seems to show that they are all less satisfactory than *M. Glaziovii* for plantation purposes. Very large yields have been attributed to young trees of the Maniçoba species in Brazil. The tapping of *M. dichotoma* is usually carried out with the knife. When the latex is collected in vessels it begins at once to coagulate. It is then moulded with the hands into balls, which are usually pressed between rollers and afterwards thoroughly dried. *Manihot piauhyensis* and *M. heptaphylla*, on the other hand, are tapped at the base of the stem, and the latex allowed to flow upon the ground.

Castilloa.

Castilloa elastica is the source of the Ule rubber of Central America and of the Caucho rubber of Peru. The species extends from the South of Mexico to the North of Peru, where it is separated from the territory

of *Hevea* by the watershed of the Andes. In Panama
a different species, *Castilloa Markhamiana*, occurs, which
is considered less satisfactory as a source of rubber. It
is probable that it is this species which has been introduced into the East. This may account for the somewhat
poor results which have been obtained with *Castilloa* in
British Colonies.

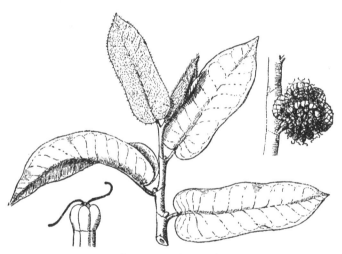

Fig. 3. *Castilloa elastica.*

Castilloa has very large oval leaves, which are
arranged along both sides of special branches. The
latter fall off with the leaves, and the whole thus
simulates a compound leaf. The flowers are massed in
dense heads, and each gives rise to a single seed which
is white and comparatively small. The bark is very

hard throughout, and offers considerable resistance to tapping. The latex tubes are larger than those of *Hevea* or *Manihot*, and are entirely devoid of the cross partitions which characterise the early stages of the latter. Partly for this reason more rubber is obtained at a single tapping, but the trees must then be rested for a considerable time before a fresh supply becomes available. The trees occur naturally in small groups along the banks of running streams from sea level to an altitude of 1500 feet, but they do not grow well in marshy soil.

Tapping is carried out by making long slanting cuts in the bark, sometimes extending right round the tree in the form of one or more spirals. The flow of latex is complete in 24 hours, and in some districts the tree is then cut down in order to obtain the whole of the available latex. The latex is acid in reaction, and coagulation is effected by the addition of the alkaline juice of certain plants or of a solution of soap. The process is sometimes carried out in rectangular pits in the ground lined with clay. The rubber thus obtained contains numerous impurities. Owing to its greater purity, the rubber which has dried on the tree fetches a higher price.

Other American Species.

Hancornia speciosa, the source of *Mangabeira* rubber, occurs in the South-West of Brazil and extends as far south as Rio de Janeiro. It grows on plateaux at

an elevation of 3000 to 5000 feet. It is a small tree
with narrow leaves and a drooping habit like that of
a weeping birch. The fruit is yellow and edible, somewhat resembling a plum. Tapping is carried out in
various ways similar to those adopted in the case of
Manihot, but it is said that all the different processes
are leading to the extinction of the plant. Coagulation
is generally effected by the addition of a solution of
alum or some other salt. The result is a wet rubber
which is not much valued in this country. The yield of
rubber is, however, said to be considerable, and the
plant appears to deserve closer attention than it has
hitherto received.

The same remark applies to some of the species of
Sapium, which grow at high elevations in Colombia and
Guiana. The Sapiums are large and hardy trees ;
nevertheless they have been largely exterminated in
their native forests by reckless tapping.

Other rubber-producing species of the Western
Hemisphere deserving of mention are *Forsterionia
gracilis* in Guiana and *F. floribunda* in Jamaica. These
are climbing plants comparable with the Landolphias of
Africa.

Guayule Rubber.

Before proceeding to describe the wild rubbers of
Africa and Asia, mention must be made of one other
American species of strikingly different habit from those
so far enumerated. This is *Parthenium argentatum,*

the Guayule rubber plant of the deserts of Northern
Mexico.   The Guayule is a small shrub, rarely attaining
a height of four feet or a diameter of more than three
inches.   The plant contains no latex, but granules of
true rubber occur scattered through the tissues, especially
those of the bark.   The plants are gathered as a whole,
and the rubber is obtained partly by a mechanical
process of grinding and partly by chemical extraction.
The exact processes employed are kept secret.   Guayule
rubber was only first placed upon the market about
1903, when the demand for rubber was beginning to
outrun the supply.   By the end of 1909 the extraction
of the rubber was said to be one of the most important
industries of Mexico.   The supply is however already
diminishing, and this source of rubber can scarcely be
regarded as a permanent one.

AFRICAN RUBBERS.

The Lagos Silk Rubber, *Funtumia elastica*, formerly
known as *Kickxia africana*, is a handsome tree somewhat resembling a coffee plant when young.   This
rubber first came into notice in the colony of Lagos
in 1894.   The trees are tapped on a herring-bone system,
and the latex collected in a vessel at the foot of the tree.
In some districts the latex is coagulated by boiling ;
this method yields an inferior rubber owing to the
damage often caused by excessive heating.   A second
method is to pour the latex into a tank excavated in the

trunk of a fallen tree. The strained milk is added from day to day until the tank is full. It is then covered with palm leaves and left for 12 to 14 days or more, according to the season, until most of the watery portions have either evaporated or sunk into the wood.

Fig. 4. *Funtumia elastica.*
A. Branch.   B. Inflorescence.   C. Flower.

The rubber obtained has a dark brown colour externally, with a paler section.

Funtumia is very susceptible to damage by intemperate tapping, and like other African rubbers, it has been largely exterminated in many districts owing to over-zealous collection. Recently plantations of this species have been established in several parts of West and Central Africa.

Although unsuitable for plantation purposes, the

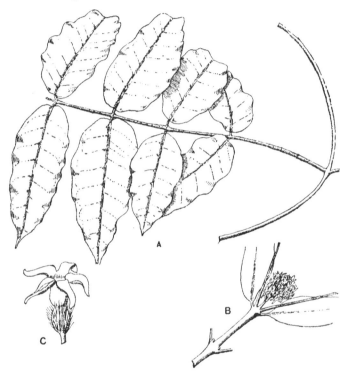

Fig. 5. *Landolphia Kirkii.*

A. Leafy twig.   B. Inflorescence.   C. Flower.

various climbing species of the genus *Landolphia* are still the chief source of wild African rubber. The

Landolphias are widely distributed over the whole of tropical Africa, extending from 16 degrees North latitude to 23 degrees South latitude. They are found in almost all the forest regions of this area, and include at least ten species of valuable rubber-yielding plants. Among the most important are *Landolphia owariensis*, which occurs throughout West Africa and the Sudan, *L. Heudelotii* in West Africa, and *L. Kirkii* and *L. Dawei* in East Africa. *Landolphia florida*, which has also an extensive range, was known as a handsome and sweet-scented flowering plant long before its commercial value was recognised.

The rubber is exported in a great variety of forms, often as small balls or sausages. These are formed by winding up the strings of rubber which dry upon the plants, when the latter are gashed with a knife.

Ficus Vogelii is another rubber plant widely distributed in West Africa and the Sudan. There have been varying reports upon its produce, and it appears to deserve further attention.

In all the above-named species the source of the rubber is the stem. The so-called root rubbers of the Congo and Angola are derived from the Rhizomes of two semi-herbaceous plants *Carpodinus lanceolatus* and *Clitandra henriquesiana*. The creeping underground stems of these plants are about an inch in diameter. The natives extract the rubber by rasping and then boiling them in water.

In Madagascar rubber is obtained from other species

of *Landolphia*, and from a tree, *Mascarhenasia elastica*, which is related to *Funtumia*, and resembles it in many respects.

Fig. 6.   *Ficus elastica.*

ASIATIC RUBBERS.

We now come to the tree which until quite recently most people have been accustomed to regard as the indiarubber plant *par excellence*.   In its native home in Assam and Upper Burma, *Ficus elastica* grows into an enormous tree, which few people would recognise at first sight as the familiar indiarubber plant of suburban front rooms.   The adult tree develops huge buttress roots, and the leaves on old branches are only three or four inches in length. The fruits are little greenish-yellow figs, half an inch long. They form one of the principal foods of flying foxes.

In tapping, cuts are made on the aerial roots and even on the horizontal branches at intervals of a few inches. The tree has to be climbed twice, once for tapping, and again a day or two later in order to collect the rubber. The dried rubber is torn away from the cuts and rolled together into a ball. Eight or ten pounds is said to be obtainable from a tree at one tapping.

Ficus elastica also occurs in Java and Sumatra, and both here and in Assam considerable plantations of it have been established. In the Dutch East Indies, however, it is being rapidly ousted by *Hevea* as a plantation rubber.

Prior to the introduction of *Hevea*, the best rubber of the Malay peninsula was obtained from species of *Willughbeia*—large woody climbers, with stems six or eight inches in diameter. Other rubber-producing plants occur, but none of these are any longer of much economic importance. Some of the species concerned extend eastward to Siam, Cambodia and Cochin China. Among these another climber, *Parameria glandulifera*, has been described as one of the most prominent. Species of *Willughbeia* also occur in Borneo, where they possess some economic importance. In New Guinea rubber is obtained from a species of

Ficus, and in Fiji from *Alstonia plumosa.* In all these countries the development of plantations is rapidly supplanting the collection of wild rubber.

Jelutong Rubber.

This rubber, called after the name used by the natives in Borneo, is of a low type containing a very high proportion of resin. It has, however, taken a prominent place in the rubber market in consequence of the recent high prices of purer kinds. Jelutong rubber comes chiefly from Borneo and Sumatra, and is derived from large forest trees of the genera *Dyera* and *Alstonia.* These are abundant in certain districts, mainly in swampy places, and may attain a circumference of as much as 20 feet. Much damage has already been done by native methods of collection, but steps have recently been taken by the governments concerned to safeguard the life of the trees. Large factories have also been opened for producing purified rubber by the use of resin solvents. Schidrowitz states that the trees can be tapped 40 times in a year without damage, and that as much as 100 lbs. of latex can be obtained from a single tree. The method of tapping recommended is done with a gouge in the form of a wide **V.**

The export of Jelutong in 1910 is said to have exceeded 25,000 tons. The crude product contains only 12 to 14 per cent. of actual rubber. There is also

30 to 40 per cent. or more of water, and the remainder is chiefly resin. This so-called rubber may be used directly in combination with a certain proportion of a purer grade for the manufacture of inferior classes of goods, or the resin may be more or less completely extracted by the use of solvents.

CHAPTER III

RUBBER is derived from a milky liquid, known as *latex*, which occupies a special series of channels in the *cortex*, or inner bark of a number of different species of plants. The proportion of rubber contained in the latex varies greatly in different species. In many latices rubber is almost or entirely wanting, its place being taken by various resinous substances. The majority of the plants in which the latex contains a large proportion of rubber are either trees, or shrubs, or woody climbers. Before considering the origin and functions of latex, we propose to describe in very brief outline the structure and functions of those parts of a woody plant or tree which are directly or indirectly affected when the latex is removed by the operation of tapping.

The Structure and Functions of the Vegetative Organs.

What are known as the vegetative organs of a tree, as opposed to the organs specially concerned with reproduction, may be divided into leaves, stem and roots. The functions of the roots—to take the last mentioned organs first—are, firstly, to hold the tree firmly upright by anchoring it in the soil; and secondly, to absorb certain substances contained in the soil which are essential for the nourishment of the plant. Among the most important of these substances, in addition to water, are various compounds of nitrogen, phosphorus and potash.

Before these substances can be utilised as food by the different parts of the plant, it is necessary for them to be altered and in most cases combined with the still more important substance carbon, which is only obtained through the leaves. One of the most important functions of the leaves therefore is to absorb carbon from the atmosphere in the form of carbonic acid gas. The functions of the leaves however are by no means confined to absorption. In addition we may compare the leaves to so many minute kitchens or chemical laboratories, wherein the different ingredients of the food of the tree are prepared and compounded into a form in which they can be utilized and readily digested by the cells that make up the roots, stem and other organs. The energy required for these transformations is derived from the rays of the sun; and in order that the necessary chemical changes may be properly carried out, it is necessary for the leaves to be spread out in a position where they are well exposed to air and sunshine.

We may next pass on to consider the functions of the stem or trunk of the tree. The first of these is to support the leaves in a position well exposed to light

and air; the second is to provide a channel for conducting the necessary mineral substances from the roots to
the leaves in a state of very dilute solution, and also for
conducting the elaborated food supply downwards from
the leaves to the roots. The liquid which thus circulates
through the different organs of the tree is popularly
known as the sap. A third function carried out by the
trunk or stem of most plants is the storage of reserve food
materials, which are accumulated in special cells, often
in the form of starch.

As it is from the trunk of the tree that rubber is
derived in the vast majority of cases, it is necessary to
enter rather more fully into the structure and functions of
this region. The trunk of a tree is well known to consist
of two main portions—the wood and the bark, including
under the latter term the layers described by botanists
as the cortex. The cortex may be roughly defined as
the softer internal part of the bark which adjoins the
wood. If the bark is stripped from the wood, separation
takes place at an extremely soft and delicate layer of
tissue known as the *cambium*. The cambium, as we
shall see later, is the seat of growth in thickness for both
wood and bark.

Channels for the conduction of sap occur both in the
wood and in the bark, and two entirely different streams
of sap are associated with these two regions. An upward current of sap occurs in the outer part of the wood
through a series of minute vessels, in which the mineral
substances absorbed by the roots are carried to the

leaves in a state of very weak solution. The perfected food materials are carried down through definite channels, known as *phloem* tubes, in the inner part of the bark, by a downward stream of sap which is entirely independent of the upward stream. In species which produce latex there is also present in the inner bark a special system of minute vessels or tubes, which contain an emulsion of rubber and other substances. This system is closed, and is entirely separate from both the above-mentioned sets of channels, having nothing to do with either of them, although in position it is closely associated with the phloem tubes which carry the downward sap current. The upward and downward streams of sap are found in all trees, but latex tubes occur in only a comparatively small number of species.

A most important organ of the tree is the cambium. It is in this layer of very delicate cells that the growth and formation of new tissues are continually going forward. On the inner side of the cambium fresh layers, consisting of newly constructed vessels, fibres and cells, are constantly being added to the wood. On the outer side of the cambium, and consequently on the inner side of the bark, similar new additions are constantly being made to the latter. These additions to the bark provide for the increased strain on the capacity of the conducting vessels, consequent on the general growth of the tree; and replace the losses occasioned by ordinary wear and tear, or in the case of cultivated rubber trees by the tapping knife. In addition to the new channels

for the descent of sap, these additions include in certain
cases new vessels for the storage of latex.   We shall see,
however, that in the majority of latex-producing plants,
the growth of the laticiferous system is independent of
the cambium.

The laticiferous system.

The channels, which contain the latex, occur in leaves,
stem and roots.   They are already present in the young
seedling, and they may also occur in the fruits and
seeds.   In the stem and roots these passages are usually
confined to the bark.   The laticiferous channels may
arise in the bark in one or other of two distinct ways,
characteristic of different groups of plants, and the
resulting passages are distinguished as *latex tubes* and
latex vessels respectively.

Latex Tubes.

The type of channel which is characteristic of the
majority of latex-producing plants, although not of the
most important species from the point of view of
rubber-production, is known as a latex tube.   So far
as is known latex tubes are never renewed from the
cambium.   The tubes arise in the seedling in the form
of a small number of closed cells, which appear to be
capable of almost unlimited growth and extension.   As
the seedling grows, these tubes also grow and insinuate
themselves between the growing cells of the cortex,

keeping pace by branching with the increase in circumference of the stem. In *Funtumia* the whole of the adult laticiferous system is derived in this way from eight primitive cells which are already present in the embryo. How such a system is renewed, after the latex has been drawn off by tapping, does not appear to be clearly understood. As a matter of fact some months must elapse after a tree of *Funtumia* or *Castilloa* has been tapped, before a full supply of latex can be drawn off again, whereas in the case of *Hevea* and *Manihot* tapping can be carried out daily without any diminution of flow.

Latex Vessels.

In *Hevea* and *Manihot*, latex vessels are formed in an entirely different manner by the fusion of rows of cells derived from the cambium. These cells arise by the division of cambium cells in precisely the same way as any other cells of the inner bark. When first formed they are more or less brick-shaped. They are nearly square in section and their length is three or four times as great as their breadth. The particular cells which ultimately fuse to form the latex vessels are arranged in the form of a network which is generally one cell thick. The meshes of which the network is composed consist of one or two rows of cells, and are elongated in the direction of the length of the stem. The walls which intervene between the separate cells of the same network soon break down, and a more or less free passage is established throughout. The interstices of the network are occupied by the *medullary rays*, which are groups of thin-walled cells running by way of the cambium radially from the bark into the wood. The medullary rays serve in part as conducting channels for food materials between the wood and the bark, and in part for the storage of reserve food supplies. The whole laticiferous system consists of a series of such networks one within the other, extending round the stem. The space between two adjoining networks is occupied by the cells and tubes of the phloem. From the arrangement of the laticiferous vessels it will be readily understood that their contents can flow more easily in the longitudinal direction than transversely round the stem.

Hevea. Gross structure of the bark.

If we begin an examination of the bark of *Hevea* from the outside of the tree, we find first a brown or grey layer of cork, which is generally thin in young and untapped trees. The function of the cork is purely protective. Secondly, beneath the cork there occurs in healthy trees a thin dark green layer of living cells. Discolouration of this layer may be taken as indicating that the health of the tree requires attention. Thirdly comes a granular yellowish or pinkish layer of tissue, which makes up the greater part of the thickness of the bark. The granules consist of groups of thick-walled *stone cells*. Owing to the small size of these groups, they offer little obstruction to the passage of a tapping knife, but they quickly turn the edge of a razor, making the preparation of microscopic sections of the bark a matter of some difficulty. The granular layer passes insensibly into a very soft white layer, which immediately adjoins the wood. These four layers make up the bark as popularly understood. If the bark is stripped from the tree, separation takes place at the cambium, which is destroyed in the process of stripping.

Minute anatomy of Hevea bark.

The networks of laticiferous vessels in *Hevea* are seen in transverse section to be separated in the radial direction by about five rows of cells and other elements. Many of these are actually the sieve tubes and companion cells of the phloem. No radial connections have been observed between the several networks, which appear to extend completely round the stem, being thus concentric and having no communication with one another. The arrangement of the cells composing the networks is seen more clearly in longitudinal sections. In the younger networks, close to the cambium, the remains of the cell-walls may still be seen partly blocking the interior of the vessels; in the older vessels the transverse and most of the longitudinal walls between adjoining cells become absorbed. Two stages in the

dissolution of the cell-walls are seen in Figs. 8 and 9.
Fig. 8 represents a tangential section of a network very
close to the cambium, whilst that shown in Fig. 9 is
older and further away.

In the seedling the nodes of the network may be
partly formed by tubular outgrowths of the cells which
meet and fuse together, thus recalling the behaviour of

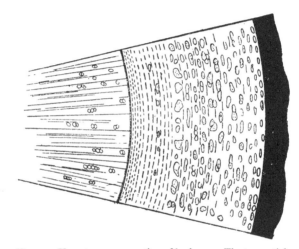

Fig. 7.   *Hevea* transverse section of bark × 5. The tangential
lines represent groups of latex vessels.

latex tubes proper.   Fig. 10 shows such a fusion in a
transverse section of the seedling.   But in the secondary
laticiferous tissue derived from the cambium there is
hardly any outgrowth of this kind.   The complete network is laid down *in situ*.

In material which has been preserved in strong

Plate IV

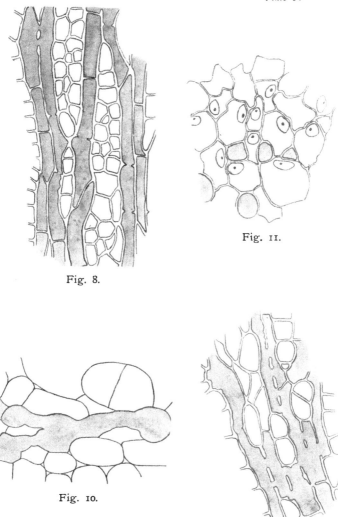

Fig. 8.

Fig. 11.

Fig. 10.

Fig. 9.

Anatomy of *Hevea*

Flemming's solution, the contents of the cells which are about to form the latex vessels are distinguishable almost immediately after they have been cut off from the cambium (Fig. 11). Their appearance and arrangement at this stage is closely comparable with that of the companion cells of the phloem. Sometimes the nuclei of the surrounding larger cells are to be observed closely pressed against the walls next to the incipient latex vessels, suggesting that the former may play some part in providing materials for the contents of the latter. The adult bark of *Manihot* has not been so closely studied as that of *Hevea*, but the manner of formation of the vessels appears to be closely similar in the two cases. For further details, and for a full account of the structure of the laticiferous system in the seedling, the reader is referred to the papers of Dr D. H. Scott.

The effects of wounding the bark.

The further remarks in this and the following chapter refer specially to *Hevea brasiliensis* unless otherwise stated, for this is the only species of which the physiology has been at all adequately studied. Even in the case of *Hevea* further observations are much needed, especially from the chemical side.

The evil effects of ringing the bark, *i.e.* severing it by a cut which penetrates to the cambium and extends right round the trunk, are primarily due to the interruption of the downward food current. Their food supply being cut off, the starvation and ultimate death of the roots is only a matter of time, if the ringing is complete and permanent. Under these circumstances the tree may linger on for several years, but all its vital functions will be greatly impeded. Such ringing may occur more or less completely as the result of injudicious or unskilful tapping. Any cut or prick which reaches the cambium cannot fail to sever a certain number of the channels through which the sap passes down to the roots; and it is probable that the weakening effect of excessive tapping is often due as much to the starvation of the roots as to the removal of the latex. Any system of tapping which involves the cutting or pricking of the whole circumference of the tree at one time is bad from this point of view. In fact it must be considered advisable never to tap more than one-third, or at the most one-half, of the total circumference of the tree at any one time.

Special stress has been laid by Fitting on the damage caused by tapping too wide an area, and the views of this observer undoubtedly carry great weight; although some authorities consider that he has taken an unduly serious view of the effects of such a process. Microscopical examination shows that the functional conducting tubes of the bark are mostly situated very close to the cambium, inside the main layers of latex vessels. Consequently the process of paring, if carried out with sufficient skill, need scarcely affect the conducting tubes. This may account for the fact that on some estates the

spiral system of tapping has been continued for years
without any perceptible injury to the trees. Nevertheless, human skill is never perfect, and any false cut, even
if it just misses the cambium, may sever some of the
conducting tubes. This danger is clearly increased
when a large part of the circumference is tapped at
one and the same time. For this reason we do not
recommend the spiral system of tapping to those com-
mencing work on new estates.

Fitting has also studied the effect of tapping in
depleting the food supplies stored up in the bark. He
found a marked reduction in the amount of starch
present in the cells around and especially immediately
below the wound, after paring had been carried out for
some time. This fact suggests that food materials are
rapidly used up in the neighbourhood of the wounded
area. It may therefore be concluded that the removal
of large quantities of latex will tend to produce partial
starvation of other parts of the tree, an effect quite
distinct from any mechanical injury.

Renewal of bark.

The bark is constantly undergoing increase in thickness owing to the activity of the cambium. As the
phloem tubes grow older their functions are given up,
and their place is taken by the younger elements within.
A considerable part of the older phloem and medullary rays becomes converted into stone cells, which

presumably serve a protective function. Active latex vessels are distributed through a large part of the area occupied by stone cells. Ultimately these also lose their functions and dry up. Finally their remains are cut off by layers of cork, and the rubber which they contain is probably lost with the dry bark which rubs off from the surface of the tree.

When the outer part of the living bark is removed in the process of paring, the cambium and young bark remaining are stimulated in a healthy tree to still more active growth, and the tapped area is said to undergo renewal. There is evidence to show that the effect of this stimulus is not confined to the area actually tapped, but that the bark of neighbouring untapped areas is also stimulated to more rapid growth, provided the rate of tapping is not excessive. The time required by the bark for renewing such a thickness as will permit of a repetition of the paring process, varies according to circumstances. Factors which affect the rate of renewal are the age of the tree, its relative size, strength and state of nutrition, and any external circumstances which modify the nutritive processes, such as climate, soil, elevation and rainfall. Renewal is also affected by the method of tapping adopted, and by the extent to which the tree has been tapped. For although moderate tapping stimulates renewal, excessive tapping hinders this process, and the more heavily a tree is tapped—after a certain rate has been reached—the longer will be the period which must elapse before tapping can be

repeated. It follows that there must be a certain
optimum rate of tapping for any given tree. Experience alone can determine this rate. A sound judgment
upon this point is perhaps the most important item in
the equipment of the successful rubber planter. Some
knowledge of the experimental work which has been
done by others, will, however, be found useful in setting
out to acquire the necessary experience.

The Functions of Latex.

The fact that the latex vessels are entirely separate
from the channels in which the food-bearing sap is
transported, gives rise to the natural question: what is
the use of these latex vessels to the tree, and what is
the precise function of the milky emulsion which they
contain? This is a point upon which we are still
very much in the dark. From the fact that the great
majority of trees of all kinds get on perfectly well
without latex, we are driven to the conclusion that this
substance is not essential for the life of the plant. It
is certainly the case that large quantities of latex can be
removed without causing any visible injury to the health
of the tree. The fact that the removal of latex stimulates the tree to the production of large additional
quantities of this substance, having nearly the same
composition as the latex originally present, suggests
that the formation of latex cannot be regarded merely
as the excretion of a waste product. On the contrary,
the conclusion seems inevitable that latex is formed at

the expense of valuable food material, containing as it does large quantities of carbon as well as a considerable percentage of proteid material. There seems to be no reason for doubting that the removal of latex causes a sensible drain upon the supply of food available for general growth.

It has been suggested that latex vessels serve as additional channels for the transport of food; and it has been stated that the phloem is poorly developed in laticiferous plants. This is not the case in the principal kinds of rubber trees grown upon plantations, all of which possess a highly developed system of sieve tubes. Other functions which have been attributed to the latex vessels are storage of food, storage of water, storage of excretory products, or a combination of any two or more of the functions already named; or finally protection against the attacks of insects and other enemies. None of these suggestions, except the last, appear to be based on any satisfactory evidence. On the other hand, latex undoubtedly serves a protective function. Any area of bark which has been entirely depleted of latex, owing to disease or other causes, is usually soon riddled by boring insects. These seldom or never attack bark in which a good supply of latex is present. Whether this function can be regarded as a sufficient cause to account for the separate evolution of such an elaborate system in several different branches of the vegetable kingdom, is a question which we may leave for the discussion of theorists.

Composition of Hevea latex.

Fresh latex as it flows from the tree consists of a fluid emulsion which closely resembles rather thin cream in general appearance. The composition is also somewhat similar, except that the fats of the cream are replaced by a different hydrocarbon. This hydrocarbon has the empyrical formula $C_{10}H_{16}$, which is the same as that of solid rubber. In the latex, however, the rubber probably exists in a liquid form. In this case the process of coagulation is either accompanied or followed by solidification of the rubber globules. The liquid hydrocarbon existing in the latex is believed to undergo polymerisation to form true rubber by the coalescence of several comparatively simple molecules into one of a more complex character. In freshly drawn *Hevea* latex the rubber globules are, on the average, approximately one-thousandth of a millimetre in diameter, but many smaller globules also occur. In other latices the globules may be somewhat larger, but their size probably seldom exceeds one ten-thousandth part of an inch. The further study of such minute bodies is naturally a difficult matter, but it is believed that the rubber or simpler hydrocarbon is kept in emulsion by the existence of a thin skin surrounding each globule, and that this skin is either of a resinous or of a proteid nature.

According to various estimates the composition of the latex of *Hevea* is approximately as follows:

TABLE VI

Composition of Hevea Latex.

Water	50—60 per cent.
Caoutchouc	30—45
Resins	1—2
Proteids	2—3
Sugar	·5
Ash	·25

The latex from old trees usually contains a considerably larger proportion of rubber than that obtained from young trees. The composition of the latex is not much altered by moderate tapping, but if tapping is excessive the amount of rubber present may be greatly reduced.

Coagulation.

Freshly drawn latex is alkaline in reaction. The addition of a suitable amount of any kind of acid leads to coagulation and to the separation of the rubber, with most of the resins and proteids, from the watery and soluble constituents of the latex.

According to the older view of coagulation, this was supposed to be due to the bursting of the skins surrounding the globules, followed by the coalescence of the latter. Recent observations have shown that the globules may persist in the coagulated latex. This fact seems to support the view that the coagulation of *Hevea* latex by acid consists in the formation of a network of coagulated proteid, which entangles the rubber globules in its meshes and contracts upon itself. The coalescence of the globules is only effected during the subsequent pressing and working of the wet rubber. The question of the extent to which polymerisation occurs during coagulation appears to be still a disputed one.

Fresh latex also contains an oxydising enzyme which leads to a darkening in colour on prolonged exposure to the air. The enzyme may be destroyed by immersing the coagulated rubber for a few minutes in water at 80° C. and this treatment has been recommended as a commercial process in order to obtain light coloured rubber. Such treatment can be replaced to some extent by thorough washing, and completely by the use of vacuum driers, in which the rubber is rapidly dried at a fairly high temperature.

CHAPTER IV

THE PHYSIOLOGY OF LATEX (*continued*)—
TAPPING EXPERIMENTS

Introductory.

THE examination of the living latex vessels *in situ* is a matter of extreme difficulty. On the other hand, the removal of the bark from the tree is followed by immediate collapse of the vessels, by the loss of a large part of their contents and by marked changes in form. Our knowledge of the physiology of the laticiferous system is therefore almost entirely derived from a study of the behaviour of the trees during the process of tapping. Experiments in tapping *Hevea brasiliensis* were carried out by the author at Henaratgoda, Ceylon, during the years 1908—1912. These experiments were begun in consultation with Dr J. C. Willis and Mr M. Kelway Bamber, of whom the latter also collaborated in the earlier experiments.

Throughout these experiments a particular system of tapping was adopted. This was done, not because the system itself was regarded as an ideal one, but for the sake of uniformity, in order that there should be a reasonable basis for comparison between different experiments. For ordinary use on estates a somewhat different system is to be recommended. Such a system will be described in a later chapter.

The system employed in the experiments was as follows. The tree was first marked with three vertical lines. Two of these lines were placed exactly on opposite sides of the tree, and the third divided the bark on one side between the two first lines into equal areas. From a point on the middle line rather more than a foot above the level of the ground, two lines were ruled slanting upwards at an angle of 45° to meet the lateral lines, thus marking out a large **V** on the surface of the tree. Above the first **V** other **V**'s were traced at intervals of a foot. The number of **V**'s tapped simultaneously was three in the majority of the experiments. In other experiments, designed for the study of special points, six **V**'s were tapped at once on the same side of the tree; but where no statement is made to the contrary, it is to be understood that the number of **V**'s under operation at any one time was confined to three.

In the positions indicated by the **V**'s thus traced, grooves were cut in the bark to a suitable depth for tapping the laticiferous tissue without damage to the cambium. The points of the **V**'s were joined by a shallower vertical groove serving to conduct the latex to a cup placed at the foot of the tree to receive it. The above processes constitute the first day's tapping. Subsequent tappings consist in removing a thin shaving of bark from the lower side of each **V**, in order to reopen the latex vessels which become plugged with coagulated rubber during the intervals between successive tappings. By this process the whole of the outer bark between the **V**'s is gradually removed in the form of thin shavings. When this has been done the tapping of the first area of bark is said to be completed. In the majority of experiments tapping was then continued by the similar treatment of the corresponding area of bark on the opposite side of the tree, and when this in turn was finished a similar area was tapped immediately above the first area. Similar tapping of a fourth area immediately above the second area completed the treatment of as much bark as could conveniently be reached from the level of the ground. In the more prolonged experiments tapping was then continued on the renewed bark of the first area[1].

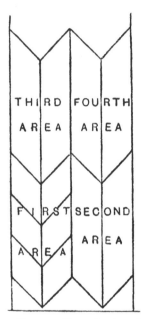

Fig. 12. Shows the tapped areas as they would appear on the bark if removed from the tree and spread out flat.

[1] During the first three years tapping in the main experiment with 70 trees, the spur pricker was used in combination with paring. This

Further details with regard to tapping will be found
in Chapter VI; but it may be pointed out here that the
method described is known as an *excision* method, as
opposed to methods of *incision* in which the bark is
merely pricked with a pointed instrument or gashed
with a hatchet or chisel. The latter is the method
generally employed in collecting wild rubber, whereas
on plantations excision is the rule.

Wound Response.

If a tree of *Castilloa* or *Funtumia* is tapped, and the
wounds are reopened after an interval of a few days, or
if the bark is again tapped after a short interval in the
neighbourhood of the original cuts, little or no latex is
obtained at the second tapping. The bark is milked
almost dry at a single operation, and the latex tubes
are not completely refilled for several months[1]. In
the case of *Hevea* and *Manihot*, on the other hand,
a good yield is again obtained after an interval of only
a single day. These facts have long been known to the
collectors of wild rubber. The *Hevea* trees in the forests
of Brazil are tapped repeatedly during a single season,

method was given up about the middle of 1911, owing to the accumulation
of evidence that the use of the pricker led to damage of the trees. The
result of subsequent tapping in which the paring knife only was used, confirmed this impression and led to a general increase of yield.

The use of the pricker may be regarded as having accentuated the
damage done to the bark of the trees tapped at shorter intervals.

[1] If indeed they ever recover. It is equally likely that much of the
latex obtained at later tappings is derived from new latex tubes budded off
from the old ones.

whereas in Central America it has frequently been the custom to cut down the *Castilloa* trees in order that the whole of the available rubber may be obtained at one time.

With the establishment of plantations, opportunities for more precise observations soon arose. Among the earliest experiments dealing with "wound response" were those of Willis and Parkin, who first made use of this expression. These experiments were carried out in Ceylon and may here be recorded on account of their historical interest.

In 1897 Willis tapped a number of *Hevea* trees of about two feet mean girth at intervals of a week with the following result.

TABLE VII

Willis. Average yield per tree (dry rubber).

	Ounces		Ounces
First week	·73	Fourth week	·80
Second week	1·48	Fifth week	·67
Third week	·97	Sixth week	·52

More elaborate experiments were carried out by Parkin at Peradeniya in 1899, and the result of one of these is recorded in the following table.

TABLE VIII

Parkin (Latex, six Trees).

March 25	...	61·0 c.c.	May	1	...	253·0 c.c.
30	...	105·5		6	...	264·5
April 6	...	220·0		13	...	275·0
12	...	208·5		20	...	255·0
15	...	255·5		26	...	262·0
20	...	290·0	June	1	...	328·0
25	...	276·0		6	...	449·0

It will be seen that after 14 tappings made at intervals varying from three to seven days (average interval 5·6 days), the yield had increased by over 600 per cent. The increased yield thus recorded was ascribed to the effect of wound response.

There seems to be little doubt that the increase recorded by Parkin was an exceptional one, and that rather too much weight has been laid on the phenomena observed. It is quite possible that part of the increase was due to climatic conditions, and this view is supported by the rainfall returns during the period of the experiment, which have fortunately been preserved. They are as follows:

TABLE IX

Rainfall at Peradeniya, 1899.

	Inches			Inches
March	5·42		May	7·59
April	18·40		June	10·74

Reasons for the increase of yield on tapping.

It is possible that the immediate response to tapping is partly due to a reduction in the viscosity of the latex, which is thus enabled to flow more freely. When part of the latex is removed from the vessels by tapping, the internal pressure of the latter is reduced and the remaining latex is diluted by the entrance of sap from the surrounding cells. The pressure in the surrounding cells may also be increased owing to the irritation consequent upon wounding. It is a fact that the proportion of caoutchouc in the latex often falls off rapidly during the first few tappings. The important point to be observed, however, is that the yield shows no diminution for a considerable period in spite of continued tapping even at daily intervals. Table XI shows the yields of dry rubber from 70 trees tapped at Henaratgoda, beginning in June 1908.

The trees were taken in groups of ten, numbered from I to VII respectively, and the average interval in days between successive tappings in the case of each group was as follows:

TABLE X

Average intervals between tappings.

Group	I.	II.	III.	IV.	V.	VI.	VII.
Days	1·275	2·225	3·35	4·425	5·525	6·55	7·325

The interval aimed at was an even number of days in each case. The actual excess was due to the fact that it was not always found possible to tap on the appointed day.

The tappings at longer intervals extended of course over a longer period of time than those at shorter intervals, consequently the climatic conditions are not identical in the case of the results recorded in different columns of Table XI. Nevertheless it will be observed that the variations in yield throughout the whole experiment were comparatively small.

In other experiments a larger initial increase in yield was observed. In such cases the results resemble

TABLE XI. *First* 40 *Tappings.   Total Rubber from* 10 *Trees tapped at intervals of* 1—7 *Days* (*nominal*).

(Weights in grammes; circumference in inches.)

No. of Tapping	I.	II.	III.	IV.	V.	VI.	VII.
			Total Circumference of each Ten Trees.				
	362	*359*	*313*	*366*	*336*	*402*	*396*
1 ...	194	143	155	149	117	135	114
2 ...	159	130	194	237	146	175	169
3 ...	224	185	197	140	146	225	188
4 ...	231	148	138	235	158	220	194
5 ...	235	133	197	190	141	241	172
6 ...	168	174	152	190	174	245	141
7 ...	194	184	189	150	190	182	142
8 ...	202	170	183	143	155	183	126
9 ...	197	160	169	156	152	209	166
10 ...	167	110	177	139	161	187	150
11 ...	174	159	161	151	166	194	160
12 ...	196	170	141	155	140	208	171
13 ...	196	140	151	150	163	223	219
14 ...	202	144	151	171	181	203	110
15 ...	179	188	158	134	149	243	169
16 ...	144	133	150	179	185	167	153
17 ...	145	153	137	147	169	209	192
18 ...	156	140	153	196	177	188	186
19 ...	87	133	136	134	176	291	154
20 ...	116	103	158	202	170	237	184
21 ...	129	123	154	180	177	208	156
22 ...	105	110	127	141	190	203	148
23 ...	98	97	153	170	180	292	156
24 ...	145	106	150	148	200	230	189
25 ...	128	82	151	171	163	188	157
26 ...	105	92	146	166	192	198	154
27 ...	134	102	110	168	251	205	160
28 ...	133	112	180	151	186	197	155
29 ...	[1]133	111	66	167	165	208	141
30 ...	126	98	211	148	159	181	133
31 ...	114	67	159	152	138	148	154
32 ...	98	86	140	159	163	190	143
33 ...	102	59	137	181	207	160	128
34 ...	94	98	139	151	157	174	115
35 ...	120	109	131	148	161	130	113
36 ...	92	132	147	120	158	158	98
37 ...	80	113	90	141	158	141	130
38 ...	85	105	152	160	184	139	92
39 ...	104	118	153	160	155	106	88
40 ...	123	65	181	164	147	104	81
Total	5,914	4,885	6,124	6,494	6,707	7,725	5,953
Total divided by circumference	16·3	13·6	19·5	17·7	20·0	19·2	15·0

[1] An interval of 11 days between the 28th and 29th tapping of row I., and between the 14th and 15th tapping of row II.

those of Parkin more closely. It must be remembered also that the method of tapping employed was entirely different from that of Parkin, who used an incision method pure and simple, and did not reopen the same wounds. The result of all such experiments may be summarised as follows. With any moderate method of tapping, carried out at nearly equal intervals, which may vary in length from 24 hours up to 10 days, the yield per tapping rises rapidly to a point at which it is subsequently maintained, subject to certain variations the nature of which will be considered later on.

Duration of yield.

With moderate and careful tapping no limit can at present be set to the period during which a similar yield will continue to be obtained from *Hevea*. By the end of April 1912, the tapping of the same seventy trees had been continued without intermission for nearly four years. The annual yields obtained are summarised in the following table:

TABLE XII

Yields calculated to lbs. of dry rubber per acre.

Year	Row						
	I.	II.	III.	IV.	V.	VI.	VII.
1908 (7 months)...	950	590	485	390	335	345	240
1909 ...	890	600	480	380	360	340	265
1910 ...	900	540	380	380	370	300	255
1911 ...	(350[1])	700	500	620	580	500	370
1912 (4 months)...	—	180	145	220	240	220	165
Average annual yield for 4 years ...	770	650	500	500	470	425	325

[1] For 4 months.

The result is here calculated in the form of the weight of dry rubber which would be obtained from an acre planted at the same distance as the actual trees under experiment. After continuous tapping for nearly four years the trees, especially those tapped at the longer intervals, were yielding rubber in considerably larger quantities than the average amount for the first year. It must be pointed out that the trees in question were upwards of 25 years old, and had not been regularly tapped before the experiment began. They are also very closely planted, namely at a distance of only 12 × 12 feet.

Here we appear to have evidence of a very real and prolonged response on the part of the trees to certain stimulating causes. Among these, two definite stimuli may probably be distinguished, firstly the removal of the latex, and secondly the irritation due to the wounding of the bark. In the case of young and vigorously growing trees a further reason for the increase in yield is apparent, namely the rapid increase in the total volume of the bark. In the case of the trees used in the foregoing experiment, the average increase in girth was less than one inch per annum, so that the increase in volume of the bark was very small in comparison with the amount present at the beginning of tapping. In trees growing under more favourable conditions the increase is often three or four times as great. In the case of young trees planted at wide intervals no limit can at present be set to the period during which a gradual increase in yield may be shown, so long as the tapping is not severe enough to cause definite injury.

Relation of yield to volume of bark.

The evidence already given is sufficient to show that the amount of latex which can be removed from a *Hevea* tree in one year, must be very large in comparison with the quantity which was present in the vessels prior to the commencement of tapping. In further illustration of this point we may take the case of an exceptionally large and vigorous tree, the yield from which was separately recorded for a considerable period.

This particular tree is one of those originally planted at Henaratgoda in 1876. Its girth at three feet from the ground was 102 inches in December 1908 when the experiment was begun; and it had increased to 115 inches in December 1911, at which date the available records come to an end. The tree therefore continued its normal rate of growth of upwards of three inches per annum throughout the whole course of the experiment.

The tree was tapped daily on the three-V system described above, and in the space of a little more than two years, four similar areas of bark had been completely tapped. The renewed bark on the first area was then retapped, and was nearly completed at the end of December 1911. During this last period tapping was carried out daily during alternate months, with monthly intervals of rest. The yield of dry rubber was as follows:

TABLE XIII

Yield of dry rubber in three years from a single tree.

Section	No. of Tappings	Total Rubber produced, grammes	Average per Tapping, grammes
I.	153	19,794	130
II.	185	19,744	107
III.	137	15,832	116
IV.	125	22,827	183
I. on renewed bark not completed	150	30,508	203

In three years the total yield was nearly 240 lbs. of dry rubber. This was contained in about 70 gallons of latex, or nearly 20,000 cubic inches. The total area of the bark tapped during this period was about 10,000 square inches. The thickness of the laticiferous bark was about half an inch, so that the volume of the bark actually subjected to tapping was about 5000 cubic inches. It is by no means an easy matter to estimate the volume of latex vessels contained in a given volume of bark, owing to the marked shrinkage of the former when the bark is removed from the tree. It will be safe however to assume that the total volume of the latex vessels is not more than one-tenth of the total volume of the bark, and most writers upon the subject have given a considerably lower estimate. We have therefore obtained in three years 20,000 cubic inches of latex by tapping an area of bark which did not contain more than 500 cubic inches of latex at the beginning of the experiment. The problem before us is to account for the remaining 19,500 cubic inches of latex.

The tree is about 80 feet high, and at about ten feet
from the ground it divides into three main branches
each about four feet in girth.   In comparison with the
girth at the base of the trunk, the crown of the tree is
by no means extensive.   In fact the volume of the
bark per foot of altitude probably falls off rather than
increases as we pass upwards, since although the total
circumference of the branch system increases, the thickness of the bark rapidly diminishes.   We may therefore
assume with some degree of confidence that the total
volume of the laticiferous system of the whole tree does
not exceed 5000 cubic inches, and is probably very
much less.   Even if this very liberal estimate be adopted,
the whole system must have been emptied twice over
in the space of three years.   At the end of this period
the vessels were still full of latex, and the freedom of
flow had increased rather than diminished.

Origin of Latex.

What is the source of this very large supply?   It has
been suggested that the latex may be manufactured in
the leaves and pass down the tree to the wounded area.
Now *Hevea* is a deciduous tree and in Ceylon drops the
whole of its leaves in February or March.   There may
be an interval of nearly a month between the loss of
the old leaves and the time when the young leaves
become fully functional.   Hence, if the leaves represented
the chief source of the latex supply, we should expect
the yield of latex to be reduced almost to nothing
shortly after the fall of the leaves, if tapping be carried

on without intermission. We shall presently see that there is some falling off in yield about this period, but that the reduction is not nearly so great as would be expected on the theory of the origin of the latex in the leaves. Moreover the composition of the latex in the leaves differs considerably from that in the stem.

During the experiment the tree was steadily increasing in girth, and new latex vessels were constantly being added to the bark by the activity of the cambium. The amount of bark tissue thus added in three years over the area actually tapped was nearly equal to the amount removed, but in other parts of the tree much less. The new latex vessels contained in the renewed bark are therefore only responsible for a comparatively small proportion of the latex obtained. The greater part of the latex still unaccounted for can only have been produced in one way, and that is by the active secretion of latex in existing laticiferous tissue. In fact we arrive at the important conclusion—not by any means universally accepted—that the laticiferous tissue of the bark is an organ for the manufacture of latex as well as for the storage of latex.

There is evidence that this process of manufacture is specially active in the immediate neighbourhood of the wounded area. In an experiment carried out on 29 trees of an average girth of 26 inches, the whole of the outer bark on one side was removed to a height of six feet by tapping on a six-V system for a period of only six months. Tapping was immediately continued by a similar system on the other half of the tree. The cuts were twelve inches apart, and, after the first day's paring had been carried out, each further cut except the lowest was apparently draining an almost isolated patch of bark, of an average area of 216 square inches. In this case the average thickness of the laticiferous bark was not more than a quarter of an inch, and on our previous assumption the average volume of the latex vessels contained in one of the isolated patches of bark, could not have been more than $5\frac{1}{2}$ cubic inches. The yield in six months from the lowest cut but one was 55 cubic inches of latex, more than 10 of which were drawn off during the first month of tapping.

There must nevertheless be a considerable movement of latex from untapped bark towards the seat of tapping. It was calculated that the largest day's yield from the old tree previously described would have drawn off the whole of the latex from a distance of at least three inches from the tapped surface. As the tubes are never completely emptied, there was probably an actual flow from at least double this distance. In the case of this particular tree the flow continued for several hours, whereas in the majority of trees the cut surface becomes blocked by coagulated rubber in less than one hour after tapping. The structure of the laticiferous system shows that movement of the latex will take place more readily in the vertical direction, but that a lateral flow is also possible. In the older networks a gradual transference of latex may take place over an extensive area.

Seasonal variation.

Some idea of the variations in the flow of latex at different times of year on the Western side of Ceylon may be obtained from the following table, which embodies the results of three years' continuous tapping of the 70 trees already described as having been tapped at different intervals by groups of ten. Tapping was carried out during the last seven months of 1908, so that by the beginning of 1909 the trees may be supposed to have been responding freely to the influence of tapping. By taking the average of three years we can eliminate to some extent the effect of different stages of tapping. This is necessary because, at least when the tapping intervals are short, the yield is greatest soon after a particular area is begun, and falls off when the bark on that area is becoming exhausted. Row I is omitted because it was only tapped for a few months during 1911.

TABLE XIV

Average yields per tapping, 1909—1911 (grammes).

	II.	III.	IV.	V.	VI.	VII.	Average II.—VII.
January	83	96	111	141	148	144	120
February	74	89	110	127	133	127	110
March	61	83	106	113	115	99	96
April	(57)	72	92	115	116	88	90
May	63	70	97	118	107	93	91
June	60	75	109	137	120	129	105
July	77	71	102	134	136	129	108
August	84	72	104	130	129	112	105
September	84	74	113	135	141	137	114
October	84	77	116	(141)	160	138	119
November	91	89	125	151	167	153	129
December	87	84	128	161	171	158	131

It appears that on the average the yield per tapping is lowest in April and May. This yield rises continuously until December and then steadily falls off again. The increase from April to December on the average of these three years, and taking the average result from tapping at a number of different intervals, does not fall far short of fifty per cent.

The seasonal variation in yield is clearly associated to some extent with the climatic conditions at different seasons, as may be seen on comparing it with the following table which shows the average monthly rainfall at Henaratgoda for the three years 1909—1911.

TABLE XV

Rainfall at Henaratgoda, Averages 1909—1911.

Month	Rainy days	Rainfall (inches)	Month	Rainy days	Rainfall (inches)
January	4	1·54	July	10	8·88
February	1	·61	August	9	7·04
March	6	3·95	September	5	2·91
April	6	4·43	October	16	24·86
May	10	6·69	November	16	11·81
June	12	9·01	December	9	7·99

It will be observed, on comparing Tables XIV and XV, that the season of highest yield follows shortly after the season of greatest rainfall. In countries subject to a prolonged dry season the variations in yield are much more marked. The following table shows the monthly crops of dry rubber harvested during 1912 on a well-known property in South India.

TABLE XVI

Stagbrook Rubber Co. Crop of dry rubber in lbs.

Month	Crop	Month	Crop
January	2345	July	2256
February	nil	August	2809
March	nil	September	5871
April	nil	October	9087
May	1205	November	14088
June	2514	December	18000

More than half the total crop was harvested in the last two months of the year.

Some relation can also be traced between the yield and the rainfall at different seasons in any given year. It seems clear however that the variation at Henaratgoda is also affected by other factors. The highest yields are obtained in November, December and January, just before the time of leaf fall. There is then a rapid falling off, and the yield remains low until August, that is to say during the whole time of the formation and ripening of the fruits. This is what would naturally be expected on general principles, although the differences are perhaps less marked than we might have been led to anticipate. Whatever the function of rubber may be, there can be no doubt that its formation constitutes a tax on the food supplies of the tree, whilst the latex removed in tapping contains other materials of possible nutritive value. We should expect the available supplies of food material to be greatest just before the fall of the leaves, since these are engaged up to the last in

food production, and give up part of their organic
contents to the tree before they drop to the ground.
In April and May, on the other hand, the formation of
fresh leaves, flowers and fruits has successively drained
the resources of the tree, so that these must be at a
comparatively low ebb. We find, accordingly, that the
percentage of caoutchouc is highest in December and
January, and lowest in May and June.

Variation in yield of individual trees.

It is well known that rubber trees possess a marked
individuality as regards the amount of latex which can
be drawn from them. Tapping coolies, if left to themselves, soon discover these differences, and confine their
attention to the best-yielding trees. The differences are
often great, and to a large extent independent of the
girth of the trees. Among a group of 29 trees of
uniform age tapped daily, the highest and lowest
average yields for the first 30 tappings were respectively
166 and 8 cubic centimetres. The circumferences of
these two trees were 52 and 32 inches respectively, and
they were not the largest and smallest trees of the
group. The yield per inch of bark removed was in the
ratio of 317 to 25, or more than 12 to 1. Such
differences afford good grounds for anticipating that
markedly increased yields could be got from plantations
derived from the seed of trees selected for their high
yielding capacity. As we shall point out at greater

length in Chapter V, the selection of seed-bearers may be expected to play an important part in the future development of the rubber-planting industry.

Tapping Intervals.

The increased yield associated with wound response is clearly manifested when the interval between successive tappings is anything between one and ten days[1]. The longest interval over which this effect can persist is not known, and would doubtless be found to vary in the case of different trees. It was observed, however, in the case of a very large tree which was tapped daily, that, when daily tapping was renewed after an interval of a month during which tapping had been suspended, the first day's yield was much smaller than the average, and was followed by a rapid increase on the days immediately following. A longer interval than a week between successive tappings is not likely to enter into the calculations of the practical planter. Which is the most profitable among intervals shorter than seven days, *i.e.* which will lead to the largest permanent yield, is a question that requires to be made the subject of further discussion and experiment.

The view generally accepted at the present time is that, with a view to obtaining the largest possible yield without permanent damage to the tree, tapping should proceed at such a rate that the paring of the whole area

---

[1] Except perhaps in the case of very young trees.

of bark accessible from the ground occupies not less than four years. Under these circumstances it will not be necessary to touch the oldest renewed bark until the beginning of the fifth year. The system of tapping is generally so arranged that this object is achieved by tapping either daily or on alternate days. There is evidence however that, *in the case of old trees closely planted*—and close planting is the rule on the older plantations—a better result can be obtained by increasing the interval between successive tappings.

In this connection reference may be made to Table XII, which shows the annual yields from 70 old and closely planted trees, divided into groups of 10, which were tapped at different intervals. The rate of tapping of the different groups was such that the original bark would be exhausted at approximately the following rates.

TABLE XVII

Time occupied in tapping original bark.

Row	I.	II.	III.	IV.	V.	VI.	VII.
Years	2	4	6	7	8	9	10

Although Group I gave the highest total yield, the bark was so much injured by the rapid tapping that no further extraction of latex was possible after May, 1911, and it is anticipated that some years must elapse before tapping can be resumed. All the other groups show increased yields in 1911 and 1912, but the relative increase is greater in the case of the groups tapped at longer intervals. In illustration of this difference, details are here given of the yields during the first four months of 1912.

TABLE XVIII

Yield of dry rubber, January to April, 1912.

	Row					
	II.	III.	IV.	V.	VI.	VII.
Number of tappings	42	32	24	20	16	14
Yield of dry rubber (grammes) ...	3,164	2,216	3,873	4,072	3,765	2,862
Yield per tapping (grammes) ...	75	70	161	203	235	202

The above figures may be compared with those in Table XI. A more convenient comparison is given however in the following table.

TABLE XIX

Yields in grammes per inch of circumference.

Row	Average tapping interval (days)	Average yield per tapping per inch of bark, 1st 40 tappings (grms.)	Average yield per tapping per inch Jan.—Apr., 1912 (grammes)	Difference
II.	2·6	·34	·21	− ·13
III.	3·9	·49	·22	− ·27
IV.	5·1	·44	·44	0
V.	6·5	·50	·60	·10
VI.	7·8	·48	·58	·10
VII.	9·0	·37	·51	·14

It is to be observed that January to April is usually a period of comparatively low yield, whilst the first 40 tappings of groups V, VI and VII extended over the whole of the period of highest yield. The increase in the later yields per tapping is therefore certainly not exaggerated by the conditions of the experiment. Under these conditions, after the tapping has been in progress for three and a half years, it appears that the longer the interval between successive tappings—up to an interval of about a week—the greater is the yield per tapping.

If we consider the total yields of rubber per month, this yield is greatest at first from the trees tapped at more frequent intervals. The relative yield from the trees tapped at longer intervals however gradually increases. After three and a half years' continuous tapping, the yield from trees tapped once a week may become as great as or greater than that from trees tapped at any shorter interval. A discussion of the practical importance of these conclusions may be reserved for a later page.

Overtapping.

The results just recorded naturally lead on to a discussion of the amount of tapping which can be performed without injury to the tree. Overtapping may be considered either in relation to the excision of bark or to the removal of latex. It is usual to discuss only the former kind of loss in this connection, but there is no doubt that a tree can be overtapped by pricking only, without any removal of bark tissue. In many systems of pricking the damage done to the bark is at least as great as in the case of paring, and with the best possible method of incision some damage is inevitable. Removal of latex without injury to the bark is in fact impossible, but even if this were possible there would still be a limit to the amount of latex which could be extracted without injury to the tree. For the manufacture of latex necessarily uses up a certain amount of food and energy, and the supply of these in the tree is not unlimited.

The problem of the physiological effect of paring upon the tree is therefore a complicated one. In addition to the rate of removal of the bark, both the amount of latex taken from the tree and the frequency of this extraction have to be considered. In the experiment with seven groups of trees described above, Group I and some of the trees of Group II may be said to have been overtapped, because at the time when the whole of the outer bark had been removed up to the greatest height convenient for tapping, the renewal of the first area tapped was still imperfect. This result was no doubt partly due to defects in tapping, but with the most perfect tapping there must be a limit to the possible rate of removal of the bark which will permit of proper renewal.

Let us assume a system by which the whole of the outer bark is removed to a certain height in a given period. Then if at the end of this period the bark of the area first tapped has not renewed sufficiently to allow of a second tapping, it is clear that the proper period for renewal has been underestimated. From the point of view of bark-removal the trees have been over-tapped. This is the criterion of overtapping which is most often adopted in practice. It would probably be better however to extend the idea of overtapping somewhat further, and to recognise that it is possible for a tree to be overtapped owing to the excessive removal of latex, although there may be a considerable area of untapped bark still available for tapping.

In the case of young growing trees planted with plenty of space, the yield of latex may be expected to increase steadily from year to year, so long as the bark is preserved in good condition. The ideal rate of tapping may be defined as that rate which is associated with the greatest increase in yield as time goes on. Any quicker rate may consequently be regarded as involving overtapping from the point of view of latex removal. In the case of the old trees at Henaratgoda, the rate of tapping associated with the greatest increase in yield proved to be considerably slower than the rate required for a degree of bark renewal which would generally be regarded as complete.

Yield at different levels of the trunk.

The bark is thickest near the base of the tree, and it is here that the greatest flow of latex is obtained. On 29 trees tapping was carried on by six V cuts placed each one foot below the next, the lowest cut being one foot above ground level. Continuous paring for nearly six months produced the following yields of latex from the different cuts.

TABLE XX

Yields of latex at different distances from the ground.

Height above ground level, feet	Yield of latex, cubic centimetres
6	16,607
5	18,281
4	18,773
3	22,353
2	25,998
1	67,942

The lowest cut of all gives a very much larger yield than any of the others, owing to the fact that it drains a larger area of bark, and one which is in free communication with the covering of the roots. The five upper cuts are more nearly comparable with one another. These show a steady slight increase in yield in passing down the stem. The latex from the lower cuts also contained a slightly higher percentage of rubber than that from the upper cuts.

Incidentally these figures afford further evidence in support of the conclusion that the bulk of the latex obtained is manufactured locally. If the latex passed down in large quantities from the upper part of the trunk and then flowed laterally into the tapping area, we should expect a larger flow from the topmost cut than from the cuts immediately below it. As a matter of fact the reverse is the case.

The yield of latex at six feet from the ground was rather more than three-fifths of the yield at two feet in the case of this particular experiment. The only reason for confining tapping to the lowest six feet of the trunk is therefore one of convenience. Profitable yields could doubtless be obtained at a considerably greater altitude, but this would entail great loss of time in tapping, owing to the necessity of climbing the tree.

Effect of tapping on the composition of the latex.

The increase in flow associated with wound response is accompanied by a reduction in the concentration of the latex. Hence in terms of latex the increases in yield during the early stages of tapping would be larger than those shown in the various tables in the present chapter, which refer to yields of dry rubber. Table XXI illustrates this fact. This table is the counterpart of Table XI, except that the entries refer to percentages of rubber in the latex, instead of to weights of dry rubber. Table XXI further shows that the falling off in concentration is much more rapid when the tapping intervals are small than when they are larger. After a certain period, which is longer when the intervals are longer, a more or less permanent concentration is reached, which persists for an indefinite period, but is also subject, as in the case of the yield of dry rubber, to seasonal variation. This more or less permanent concentration is considerably higher in the case of the trees tapped at longer intervals.

TABLE XXI

First 40 Tappings. Percentage of Rubber in Latex. Average of 10 Trees in each case tapped at intervals of 1—7 Days.

No. of Tapping	I.	II.	III.	IV.	V.	VI.	VII.
	Total Circumference of each Ten Trees.						
	362	359	313	366	336	402	396
1	49·7	50·0	46·2	50·0	?	?	53·5
2	49·4	45·0	40·0	39·4	43·4	43·1	41·0
3	46·0	41·2	40·0	?	37·6	41·8	36·8
4	38·9	40·4	37·4	36·5	?	39·6	35·9
5	39·0	36·1	34·9	34·3	?	39·0	?
6	36·8	38·2	?	35·7	36·7	39·2	?
7	37·6	44·4	31·7	35·2	36·8	41·3	37·7
8	36·5	42·2	34·8	36·3	39·3	41·5	36·7
9	35·2	?	32·5	36·2	38·5	40·6	36·1
10	35·2	37·9	35·0	36·8	37·2	44·4	37·7
11	34·3	?	33·4	36·0	38·9	42·5	38·6
12	32·1	30·8	34·9	38·8	35·0	41·2	40·2
13	32·8	31·7	35·6	39·5	37·7	41·3	33·4
14	30·6	?	32·6	36·3	34·8	36·3	39·9
15	29·5	29·4	34·6	41·0	?	40·4	37·0
16	28·8	25·5	32·8	37·6	39·9	38·6	41·0
17	27·9	30·3	35·2	?	40·2	40·0	36·6
18	27·5	31·4	33·5	39·2	38·1	38·0	38·0
19	26·3	31·1	31·7	?	37·9	40·0	37·5
20	27·2	?	32·7	37·6	38·6	43·3	42·0
21	26·3	30·6	32·6	38·8	39·0	40·8	38·0
22	26·4	27·2	30·1	35·5	40·0	40·8	41·7
23	25·9	30·1	33·2	39·2	38·7	42·3	41·0
24	25·9	29·4	26·0	38·5	38·3	40·4	41·5
25	26·7	29·0	33·6	38·2	40·0	42·9	38·0
26	28·5	29·0	35·0	38·2	37·0	40·3	43·9
27	28·9	27·6	?	37·3	40·0	40·3	45·3
28	30·0	32·2	35·2	39·6	42·4	43·8	42·3
29	29·8	29·5	?	41·0	39·3	42·5	39·0
30	30·2	29·4	31·1	39·4	39·0	45·5	41·5
31	30·8	29·6	33·9	39·6	40·7	42·5	39·6
32	30·2	32·8	32·3	39·5	40·5	42·2	41·4
33	26·3	?	33·8	40·4	42·9	40·7	41·0
34	27·7	30·0	35·5	38·0	39·1	43·0	39·5
35	30·4	?	34·5	41·4	42·3	46·5	36·6
36	31·2	30·2	33·0	39·7	40·8	39·0	38·4
37	29·9	31·6	34·5	40·3	39·7	39·5	37·1
38	27·2	31·0	31·3	41·5	39·5	40·2	36·1
39	28·0	?	32·3	41·0	43·5	35·8	34·3
40	28·6	29·6	34·0	43·3	40·5	37·9	?

Thus the averages of all the concentrations determined during this experiment, from 1908 to 1911, were as follows:

TABLE XXII

Percentages of rubber in latex.

	I.	II.	III.	IV.	V.	VI.	VII.
1908	32·1	33·6	35·5	39·3	38·5	40·9	40·4
1909	30·1	31·0	33·5	37·6	37·8	39·0	36·9
1910	28·3	29·5	31·2	35·5	37·4	37·4	37·7
1911	—	29·3	33·3	35·9	37·5	38·0	36·0

The seasonal variation in the concentration of the latex during prolonged tapping are similar to those observed in the total yield of dry rubber. The percentage of rubber was found to be generally highest about January and lowest about June. This fact supports the view that the yield depends upon nutrition

TABLE XXIII

Percentages of rubber. Monthly averages, 1909—1911.

	II.	III.	IV.	V.	VI.	VII.	Av.
January	34·0	34·5	39·5	40·3	41·4	40·2	38·3
February	32·8	33·9	37·9	39·2	40·2	37·6	36·9
March	28·7	34·1	36·7	37·0	37·7	37·0	35·2
April	25·7	33·7	36·8	38·4	37·6	35·9	34·7
May	26·1	30·0	34·7	35·6	35·3	31·4	32·2
June	21·2	30·2	33·9	35·5	36·2	35·9	32·1
July	28·6	29·2	31·5	36·0	34·9	36·3	32·7
August	29·4	32·6	35·7	36·6	37·1	39·0	35·1
September	30·9	36·3	36·9	38·1	37·5	37·4	36·2
October	31·8	32·6	37·2	36·3	41·1	36·6	35·9
November	31·9	32·2	36·1	37·8	38·1	38·6	35·8
December	32·6	32·9	38·6	39·9	40·9	37·4	37·0
Average	29·5	32·5	36·3	36·7	38·2	37·0	35·1

as well as upon the supply of moisture, since otherwise the concentration might be expected to be highest in the dry season and lowest in the wet. There is however no evidence of an increase in concentration during dry weather.

The concentrations here recorded were determined by dividing the weight of dry rubber by the volume of the latex, a method which gives values slightly too low throughout.

The dry rubber obtained in this experiment from the second to the sixth tapping of each group (samples A), and again after tapping had been in progress for four months (samples B) was analysed for ash, resin and organic matter. As regards ash, practically no difference was found between the product of earlier and later tappings. The proportion of resin showed a slight falling off in the product of the later tappings, especially in the case of the more frequent intervals.

The organic matter in the two series showed differences of considerable interest. In the groups tapped at shorter intervals the organic matter from the later tappings showed a distinct increase, whilst in the groups tapped at longer intervals the reverse was the case. The following purely tentative explanation was put forward to explain these differences. "If we suppose that some part of this organic matter represents the raw material from which the rubber is formed, then, in the case of daily tapping we may suppose that the drain upon the resources of the tree is at first greater than it can cope with, but that later the bark reacts more perfectly to the stimulus of tapping, and produces the raw material faster than the cells can convert it into caoutchouc. In the case of tapping at longer intervals, on the other hand, we may suppose that the stimulus to manufacture rubber leads to an accumulation of this material in the latex tubes, which is only

TABLE XXIV

Analyses of rubber from earlier tappings.

Sample No.	Resin	Chemical test		Ash
		Organic matter	Indiarubber by difference	
	Per cent.	Per cent.	Per cent.	Per cent.
1 A	2·04	0·84	96·79	0·33
2 A	2·73	0·83	96·21	0·23
3 A	1·92	0·94	96·87	0·27
4 A	2·16	0·86	96·67	0·31
5 A	1·75	0·93	97·01	0·31
6 A	2·60	1·11	96·06	0·23
7 A	2·19	1·09	96·49	0·23

Analyses of rubber from later tappings.

Sample No.	Resin	Chemical test		Ash
		Organic matter	Indiarubber by difference	
	Per cent.	Per cent.	Per cent.	Per cent.
1 B	1·78	1·59	96·35	0·28
2 B	1·84	1·43	96·43	0·30
3 B	1·79	1·43	96·53	0·22
4 B	1·93	0·71	97·14	0·22
5 B	1·67	0·52	97·67	0·14
6 B	2·40	0·44	96·94	0·22
7 B	2·19	0·63	96·95	0·23

partly removed by the *e.g.* weekly opening of the vessels, and that this concentration...inhibits the further production of raw material. In other words, tapping stimulates the tree both to the production of the raw materials for rubber and to the conversion of these into rubber. With frequent tapping the raw materials are produced faster than they can be converted; with less frequent tapping the reverse is the case[1]." It is scarcely necessary to point out that the whole subject requires further investigation from the chemical standpoint. The actual figures obtained in the analyses are given in Table XXIV.

General considerations affecting yields.

Given a uniform and moderate system of tapping, the yield from a *Hevea* tree should increase steadily with its age and girth. Too drastic tapping will certainly reduce the rate of increase of yield, and may even lead to a falling off in the absolute yield of latex and of rubber. On the other hand, there is reason to believe that the yield after a given number of years will be greater from a tree which has been regularly tapped in moderation than from one which has been left un-tapped. It should be observed, however, that the definite experimental evidence upon this point is very limited. Assuming that moderate tapping stimulates the tree to increased activity in the production of latex,

[1] Circulars R. B. G. Peradeniya, Vol. v. No. 16.

it should be the aim of the planter to discover just that rate of tapping which will lead to the most rapid increase of yield under the conditions affecting each particular field.

Just as in the case of many other physiological functions, the yield of latex at any given age would appear to be controlled by a number of limiting factors. The system and rate of tapping represent one such factor. Others are the available space, soil and water supply, and climatic conditions generally. These conditions severally react upon what is probably the most important factor of all, namely the individual character of the tree.

The rate of increase of yield will be reduced if the trees are planted too closely. But the most profitable number of trees which can be borne by any given acre at any given age, is a problem which can only be solved by prolonged trial and experience. When the soil is rich, more space is required for the full development of the trees than when the soil is poor. Similarly, more space is required at any given age at low than at high elevations.

The condition of the water supply seems to have an even more immediate effect upon the yield of latex. Where the water level is close to the surface of the soil, as in parts of the Federated Malay States, large yields are obtained at an early age. When the water level is deep below the surface, the roots may not penetrate to the requisite depth for some years, and during this period the yield is more dependent upon the seasonal rainfall. From old trees which have penetrated to the water level, larger and more continuous yields are obtained. This seems to be the case with the old trees at Henaratgoda, where the water level is upwards of 25 feet below the surface.

Resting Periods.

A question which is frequently asked concerns the necessity or otherwise of resting periods. This also is a question upon which definite experimental work is required before a positive conclusion can be arrived at. It has been recommended that tapping operations should be suspended during the period of leaf-fall, and this advice seems to be based upon reasonable grounds. In climates such as those of South India and the north-east side of Ceylon, the yield is found to be reduced almost to nothing during the prolonged dry season. In these climates the year is divided of necessity into a season of tapping and a season of rest.

From the available evidence it seems probable that in countries not subject to periods of prolonged drought continuous moderate tapping is likely to be more profitable than alternate periods of rest and severe tapping. The continuous method is also distinctly more convenient in connection with the working of a labour force. The theoretical advantage of the continuous method is associated with the habit which the tree seems to acquire of manufacturing a steady supply of latex.

The effects of tapping on the tree.

Moderate tapping seems to encourage not only increased latex formation but also an increased rate of growth in thickness over the area tapped. It seems probable that this increase must take place at the expense of the growth of other parts of the tree. So long as the only recognisable effect of this kind is a reduction in the quantity of seed produced, no harm is done from the planter's point of view. It may even be suggested that artificial removal of the young fruits might prove profitable, if it could be carried out at a small expense, as leading to the conservation of food supplies in the tree.

Severe tapping has a reverse effect in every way. If the trees are tapped to excess, growth is checked, and in particular the proper renewal of the bark is interfered with. The latex moreover becomes poor in quality and contains a smaller percentage of caoutchouc. Most serious of all is the effect upon the general health of the plant. The tree may be so weakened that it is unable to withstand the attacks of fungus diseases, which would not have been able to gain a footing if the tree had been preserved in a condition of perfect health. Canker and similar diseases seldom attack the trees unless the latter are either overtapped or very closely planted.

Summary.

In *Hevea brasiliensis* as grown under plantation conditions, repeated tapping on a moderate system at intervals varying from one to ten days leads to an immediate steady increase in the yield of latex and rubber obtained at each tapping. In the case of young and vigorous trees, this increase may continue for an indefinite period, subject to certain seasonal variations. The increase is found to take place even with daily tapping if the system adopted is such as to allow four years for the renewal of the bark. In the case of old trees, closely planted, after tapping has continued for some time, the yield from tapping at longer intervals increases relatively to the yield from tapping at shorter intervals.

This fact may be regarded as one aspect of the more general phenomenon that moderate tapping leads to a steady increase in yield, whilst overtapping ultimately leads to a relative falling off in yield. The amount of excision which constitutes overtapping depends upon the conditions and upon the individuality of the particular tree.

The bulk of latex which can be extracted in a year is probably often quite as great as the total volume which all the latex tubes of the tree could contain at any one time. Although the latex can pass gradually from one part of the tree to another, especially in the vertical direction, the greatest part of the latex removed is probably secreted at no great distance from the wounded area.

Under the comparatively uniform conditions of the climate of Western Ceylon, the yield of latex at certain seasons may be nearly twice as great as at other seasons. Seasonal differences of yield are partly determined by climatic conditions, but probably also in part by the quantity of food materials available in the tree.

The yield from different trees of the same age and girth varies greatly. A difference of 1000 per cent. is not at all uncommon.

The yield is greatest near the foot of the tree, and decreases gradually on passing upwards. In all probability the yield is roughly proportional to the volume of the bark at any given level.

The percentage of rubber in the latex shows variations similar to those displayed by the total yield. The percentage is highest during the period of highest yield. Severe tapping leads to a marked falling off in the percentage of rubber present.

It may be concluded that in practice the rate of tapping should be reduced either

(1) if the bark is being used up at a greater rate than one quarter of the available amount annually, or

(2) if the concentration of rubber in the latex falls much below thirty per cent., or

(3) if the yield of latex fails to show an increase over the amount obtained at the corresponding period of the preceding season.

CHAPTER V

Choice of situation and soil.

THE hardiness of the *Hevea* rubber tree in the different countries of its adoption has been the occasion of some surprise. In Ceylon the trees have made remarkably good growth in situations where, in the early days of their introduction to the country, botanical experts would never have supposed them capable of growing with permanent success. Anywhere in moist climates within ten degrees of the equator *Hevea* will grow, though not luxuriantly, even in rocky situations up to 2500 feet elevation. Here it may be seen entering into competition with the still more hardy tea-bush in its power of making the best of unfavourable circumstances. It will also grow in comparatively dry districts, if adequately protected from wind. The soil at Henaratgoda is remarkably poor; nevertheless, as we saw in the last chapter, phenomenal yields have been obtained there from old trees. On such a soil, however, the early growth of the trees is generally slow. In fact the appearance of the Henaratgoda trees during the first few years after their introduction, led to the erroneous conclusion that on rubber estates in Ceylon tapping could not profitably be begun until the trees had reached an age of ten years at least.

It is on rich alluvial soils in the moist low country that the tree makes its most rapid growth and gives its earliest and heaviest yields. The low-lying alluvial soils of the Malay Peninsula and the light and fertile volcanic soils of Sumatra have alike shown themselves to be admirably suited for the growth of *Hevea*. Even these soils would not be reckoned particularly rich according to the ideas of farmers in a temperate climate. Some idea of the relative proportions in which the different constituents occur in different soils may be obtained from the following table of analyses.

TABLE XXV

Analyses of Rubber Soils.

	A	B	C	D
Organic matter and combined water ...	24·080	5·12	7·8	4·20
Nitrogen	0·667	0·27	0·154	0·099
Potash	0·131	0·26	0·046	0·274
Phosphoric acid ...	0·025	0·072	0·031	0·114
Lime...	0·284	trace	0·040	2·468

A is an alluvial clay from the Malay Peninsula, analysed by Kelway Bamber.

B is a Sumatran soil analysed by Schidrowitz.

Plate V

Photo C. Northway

Hevea Rubber on Swampy Land

C is the "cabooky" soil at Henaratgoda, Ceylon, analysed by Bruce.

D is an analysis by Dyer of soil at Rothamsted upon which wheat had been grown continuously for fifty years without manure. The low percentages of organic matter and nitrogen are to be associated with the exhaustion produced by this continuous cropping. Nevertheless this soil is still much richer in mineral constituents than either of the three examples of tropical soils. Even the Malay soil, rich as it is in organic matter and nitrogen, is very poor as regards phosphoric acid, and by no means rich in potash. In spite of such deficiencies the warm moist climate of the tropics gives rise to an abundant growth of vegetation.

In short, it may be asserted that *Hevea* can be grown to a profit on almost any soil in the latitude of Ceylon up to an elevation of 2000 feet, provided the rainfall exceeds 75 inches a year, and provided a situation is chosen which is not exposed to strong winds, especially at the dry season of the year. On the other hand, the richer the soil and the lower the altitude the better, provided that on swampy lands good drainage be provided. The best test of a soil— much better than the test of chemical analysis—is the character of the vegetation growing upon it. If the growth of forest over a given area is luxuriant, other conditions being favourable, then a good growth of rubber is assured when the forest is cleared. And if it is intended to plant rubber where other agricultural products are already established, the fact of a good return from crops of any other kind may be taken as an earnest of good crops of rubber to come.

Throughout the tropics good forest land is always best for planting if it is obtainable, and it is greatly superior to grass land or land which has already been cropped. The greater part of the rubber in the Malay Peninsula has been planted where virgin forests have been cleared, and in every country this is one of the most frequent conditions. In Ceylon a great deal of rubber has also been planted through existing fields of tea, in Sumatra on old tobacco land, and in Java amongst various other products. Where such methods are adopted the cost of clearing and weeding are debited to the original crops, but the growth of the rubber must be expected to be considerably slower than on an original clearing opened for rubber alone. The latter method is therefore the one to which we shall devote the most attention.

In selecting the site for a rubber estate several other considerations besides soil and climate demand attention. Foremost among these is the question of transport. Rubber is a commodity of which the bulk is relatively small in comparison with its present value, but the weight of the produce from a large estate is by no means a negligible quantity. The conveyance of food and other necessities for the labour force involves a considerable amount of transport, and it must not be forgotten that heavy machinery will have to be introduced when the time comes for building the factory. Ready access to a railway, canal or seaport is therefore desirable, and a special feeding canal or road may have to be constructed. The development of railways has now proceeded to a considerable extent in Ceylon, Java and the Federated Malay States, as well as in India. In the low-lying coastal regions of the Malay Peninsula, water transport is extensively adopted on the canals which are necessary for purposes of drainage. Roads are being widely constructed, and the system of these in Ceylon is particularly extensive and excellent. Here too signs are not wanting that the now universal bullock cart may one day be largely replaced by the motor lorry.

Land Tenure.

In Ceylon crown land is sold outright by auction, and is subject to a reserve price. Temple lands in the Kandyan country can be leased for a period of fifty years. A large proportion of the land suitable for rubber cultivation has now been disposed of in this way, or is already in private hands. In most other rubber-producing countries the land is held on lease from the government, and the latter often retains the right to resume possession if the lessee fails to open up the land at a certain rate, or otherwise to conform to certain regulations laid down.

L.   7

Clearing.

The heavier the forest the greater will be the
expense for clearing, but in the same proportion the
growth of the rubber may be expected to be better.
The trees on the land seldom pay for working, and
much fine timber is thus wasted, only so much being
saved as is required for buildings on the estate. Clearing
is often carried out on contract. In some cases only the
undergrowth is cleared and burnt, the larger trees being
ringed and left to die. At the present day this practice
is seldom adopted. Not only is it much better to fell
and burn everything growing upon the land, but if
possible all tree stumps should be extracted and no
dead wood should be left to rot upon the ground. In
some cases no doubt the cost of such operations is
prohibitive, but the immunity from the attacks of white
ants and root diseases, which can be obtained in no
other way, is worth considerable extra expense. On
level land, moreover, it may be possible, if stumps are
extracted, materially to reduce the cost of weeding and
cultivation by the use of agricultural machinery during
the early stages of growth of the rubber. Probably the
only real objection to the use of machinery in this
connection where the conditions are suitable, lies in
the conditions of labour management. A considerable
labour force is necessary both for opening the estate
and for harvesting the rubber. In the interval between

these operations—extending over at least four years—the principal work available for the labour force is weeding. On large estates which are opened gradually, so that the first fields may be already in tapping before the whole of the original forest land is cleared, this objection does not apply. In such cases stumping and the use of agricultural machinery may be recommended wherever the lay of the land is suitable. When we consider the way in which even heavier timber is removed by far more expensive labour in newly opened districts of North America, the fact that simple stump hauling machinery has scarcely been introduced into the tropics is certainly remarkable. Its introduction may be expected to lead to considerable economy and immunity from disease.

Nurseries.

In planting up a rubber estate, nursery plants not less than twelve months old are usually employed. Hence the establishment of nurseries is one of the first operations to be undertaken. The best planting land also makes the best nursery, and a site should be chosen where the soil is as rich as possible, well drained and well sheltered from wind. The nursery should be close to the fields where the plants will ultimately be required, in order to reduce transport as much as possible. In countries subject to a prolonged dry season some artificial shade may be necessary, but the heavy shade of trees is to be avoided, chiefly on account of the damage which may be done by the drip from the leaves in wet weather. The soil should be well dug and laid out into beds of any convenient size. The most important point is the allowance of ample space for the growth of the plants. The seeds should be planted at a distance of not less than 6×6 inches, and if possible three or four times as many plants should be raised as will ultimately be required, in order that the best specimens may be selected for planting.

The impossibility of selection is the chief objection to the practice of planting seeds at stake, which is sometimes adopted but is hardly to be recommended on this account. Another method widely employed in the Federated Malay States consists in planting the seeds in baskets, which are afterwards transferred to the field bodily with the seedlings. To this method the same objection will often apply owing to the insufficient number of plants raised. Seedlings raised in baskets can be planted out earlier than others, but the advantages of the method over that of stumping as generally practised in Ceylon do not appear to be conclusive, and it is decidedly more expensive.

In planting the nursery, the seeds should be carefully laid in the ground in a horizontal position and just covered over with soil. If the seeds are planted with the long axis vertical, germination is less satisfactory and some of the seedlings are likely to be contorted and twisted. The same area should not be used a second time for a nursery unless it has been thoroughly dug over with lime, in order to destroy any fungus and insect pests which may have accumulated, and afterwards allowed to lie fallow for some months. Thorough fencing of the nursery is necessary in order to prevent the depredations of larger animals.

Seed Selection.

In the early days of rubber planting, seed for nurseries had to be taken where it could be got and selection was out of the question; but in the future it ought to be possible to lay considerable stress on the selection of seed. Among planters a marked predilection exists for the selection of seed from old trees. Probably this preference rests upon some foundation of fact, hence, other things being equal, seed should be taken from old trees so long as these have not been tapped too heavily. Much more important, however, is the selection of seed from trees of known yield. We have already had occasion to point out the marked differences which exist between individual trees in the matter of yield. It is of the utmost importance that seed for future planting should be taken from the best yielding trees. This is not such a simple matter as might appear at first sight, on account of the scattering of the ripe seeds which takes place when the fruits burst. This phenomenon places a serious difficulty in the way of the collection of seed from particular trees, except on a very small scale. On estates therefore which are going in for wide extensions and are in possession of old trees already in bearing, the following procedure may be recommended. A few acres should be set apart definitely for seed bearing, as is done for example in the case of tea. All the trees on this area should be tapped in the same way for a definite period —say for fifty tappings—and a record should be kept of the yield of each individual tree. All except the best yielding trees should then be ruthlessly cut down and the stumps extracted in order to avoid the danger of root disease. Tapping should then be stopped for the whole period during which seed is required. In this way not more than 20 per cent. of the original trees should be left standing. If the distance of planting was originally 15 × 20 feet this would leave about thirty trees to the acre. With closer planting the selection might be still more stringent. In felling, regard should be had only to yield and not to size or position. Not only would such a group of seed-bearers yield a return from the sale of seed to neighbouring estates, but when the demand for seed is over, tapping may be resumed, and it is not impossible that the yield per acre from the seed-bearers will be found to be as large as or larger than that from any part of the estate not planted from selected seed.

To those in charge of government plantations and experiment stations a further course of selection may be recommended which is scarcely practicable on individual estates. This is the method of selection by progeny which has been practised with great success in the case of many annual crops. The method consists in planting a number of definite areas, each of an acre or less, with seed taken from the best yielding individual trees selected over as wide an area as possible. If practicable the seed of at least fifty trees should be separately planted in this way. When the different plots come into bearing their yields should be compared, and a few of the best *plots* retained for seed bearing. On the selected plots a comparison of individuals should again be made by the only reliable test, namely individual tapping, and the best yielding trees only should be allowed to stand. The analogy of cinchona would lead us to believe that if this method is carried far enough a very marked increase in the average yield of latex may be obtained. In Java the proportion of alkaloid in the bark of the introduced cinchona plants has been very nearly doubled by careful selection. There is good evidence to show that the variation in the yielding capacity of *Hevea* trees is considerably greater than the normal variation in richness of cinchona bark.

Draining, Irrigation, Roads, etc.

Nurseries having been properly established, attention may next be turned to the preparation of the fields for planting. In the first place some kind of draining will

nearly always be necessary. In tropical agriculture
two totally distinct kinds of draining are to be distinguished. The first is the draining of swampy land
for the removal of superfluous water, and the second
is the cutting of transverse drains on sloping land, in
order to check the washing away of the soil, which
soon occurs when the original vegetation is cleared.

In the Federated Malay States rubber is frequently
planted on alluvial flats where the water level is only
a foot or two from the surface of the soil. In selecting
land for an estate in such a situation it is important
to remember that provision will have to be made for
carrying off the drainage water to a river or to the sea.
For this purpose canals of some size may have to be
specially constructed, and it may be necessary to acquire
the land occupied by the canal beyond the boundary of
the estate proper. The necessary canals are usually cut
by government before the sale of the land. In soil of
this kind numerous open drains are required, which may
be as much as three or four feet deep and two or three
feet wide. In some cases a drain is needed between
each row of trees, and the trees themselves are planted
on ridges formed by the materials thrown up in digging
the drains. In the Southern Province of Ceylon a
certain amount of rubber has been planted in swamps
similarly drained.

A large part of the rubber in Ceylon and other
countries is however planted on comparatively steep
hillsides, and here an entirely different system of

drainage is required. Small drains about one and a half feet in width and depth are carried across the slopes with a fall of from 1 in 15 to 1 in 25. These are made to discharge into any natural ravines or watercourses which may occur. The distance between the drains varies, according to the slope, from 100 feet down to 20 feet or less. The earth removed in digging is thrown out on the lower side of the drain. Loss of soil may be further checked by interrupting the drains with catch pits at frequent intervals. On very rocky hillsides draining may be impracticable. In this case some kind of terracing is frequently attempted by building low walls of boulders below the trees. This method is strongly to be recommended where drains cannot be cut.

Irrigation has been little practised in the cultivation of *Hevea*. In a dry climate the tree appears to suffer more from the effects of sun and wind upon its young branches and leaves than from lack of water in the soil. Experiments in growing *Hevea* under irrigation in the dry zone of Ceylon have so far been unsuccessful, though it is possible that they might have fared better if an effective wind-break could have been found.

The laying out of roads and paths leading to the different sections of the estate is an operation which should be undertaken at the same time as the tracing of the drains, in order that the necessary culverts and bridges may be provided for. Except in the case of the principal tracks, little road-making is necessary

beyond the provision of a drain on one or both sides
of the trace, and the building of suitable culverts or
paved watercourses to prevent breaching by floods.

Lining and Spacing.

The positions of the holes in which the trees will
be planted must next be marked out in lines at right
angles to a given base. It is important that these lines
should be laid out accurately in order to allow of an
uninterrupted view through the estate. If the first line
is laid out accurately, others can readily be placed by
simple measurement. A method of placing a line at
right angles to a base line is shown in the diagram. If
AC is made equal to *BC* and *AD* to *BD* the line *CD*
will be at right angles to the line *AB*.

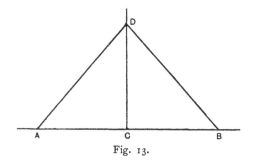

Fig. 13.

Planting Distances.

The distance at which the trees should be spaced
is a matter which has given rise to much discussion.
The distances generally favoured in the Malay Peninsula

give an average of about 150 trees to the acre, whilst in Ceylon the average number is much nearer 200 to the acre. Although different distances are suitable for different soils and situations, it is generally agreed that these numbers are too large for permanent plantations and lead to excessive crowding by about the tenth year. For future planting a distance of 15 × 15 feet may probably be regarded as the absolute minimum, giving nearly 200 trees to the acre without allowing for casualties. Such a distance must ultimately lead to crowding. On the other hand, a greater distance will almost certainly entail reduced yields from a given area during the early stages of tapping. Unfortunately there does not appear to be any conclusive evidence yet published from which a comparison can be drawn between the results of such close planting and those of wider planting, as time goes on. A consensus of opinion points to the greater value of a smaller number of larger and less crowded trees ; and it is at least probable that the crop will be obtained more cheaply from such a plantation than from one bearing a larger number of more crowded trees of the same age. Twenty trees similar to the largest tree at Henaratgoda—35 years old—would yield upwards of 1000 lbs. of rubber annually. 600 lbs. per acre is about the most that can be got from trees on similar soil planted 12 × 12 feet at 25 years of age. The large tree is however probably exceptional, whilst 12 × 12 feet is certainly too close a distance. In the present state of knowledge it may be thought that a suitable final distance is probably about 30 × 30 feet under average conditions, or 50 trees to the acre.

Where a suitable inter-crop is available, an original planting distance of 30 × 30 feet may probably be recommended. But where rubber is the only cultivation, planting at such a distance means foregoing a considerable amount of profit during the earlier stages. It is therefore not likely that such wide planting will often be adopted in practice. With closer planting the question of thinning out will have to be faced later on. This process will more properly be discussed at a somewhat later stage.

The distances mentioned in the preceding paragraphs apply to average conditions and may be somewhat increased or reduced according to circumstances. It is convenient to remember that

TABLE XXVI

a distance of feet	gives approximately trees per acre	actually trees per acre
12 × 12	300	302
15 × 15	200	194
15 × 20	150	145
20 × 20	100	109
30 × 30	50	48
35 × 35	35	35

The figures given above apply to planting on the square. By adopting the hexagonal system of planting the distance between any given tree and its nearest neighbours is somewhat increased, keeping the same number of trees per acre. This is probably an advantage during the short period when the lateral branches are just beginning to meet. The method is somewhat more troublesome than the square system and is seldom adopted. The arrangement of the trees on the two systems is shown in Fig. 14.

```
×    ×    ×              ×    ×    ×

×    ×    ×          ×    ×    ×    ×

×    ×    ×              ×    ×    ×

Square system.      Fig. 14.      Hexagonal system.
```

Holing and Planting.

After the positions of the trees have been staked out, holes must be dug for their reception. The holes should not be less than one and a half feet each way, and may conveniently be cubical. The larger the holes the better will be the general growth of the plants. They should be filled in with surface soil and allowed to settle for some time before planting. The strongest seedlings in the nurseries should be selected for planting out, and all weakly and defective plants rejected. On removal from the nursery the seedlings are frequently stumped, that is to say the whole of the green top is cut away, and the tap root is severed about 18 inches below ground level, most of the lateral roots being also cut short. Mr Tisdall tells me that it is a mistake to cut the tap root and laterals too short. This operation, which at first sight appears somewhat drastic, has great

conveniences in practice and is not theoretically objectionable. If the leaves and smaller roots were left
upon the plants, both would be found to die back
except under remarkably favourable circumstances.
Stumps can be used which have grown for two and
even three years in the nursery, whereas it would be
impossible to transplant seedlings of that age with all
their roots and leaves. A good start for the plantation
is thus assured, and there should be few failures when
proper care is exercised. The earth should be rammed
tightly round the planted seedlings or stumps. For
early planting, seedlings grown in small loose baskets
may be used, as these can be set out earlier than plants
grown in the nursery in the ordinary way. The baskets
are planted with the seedlings and are allowed to rot
in the soil. When this method is adopted it is
specially necessary to make sure that weakly plants
are rejected.

Planting should of course be carried out in wet
weather, and in most countries where rubber is grown
this presents no difficulty. Should a prolonged drought
follow shortly after planting the young plants may
require some protection. This is often best afforded
by mulching with grass and leaves, or with anything
of the kind that may be available, close round the
plants, in order to check evaporation from the soil
surrounding the roots.

Rate of Growth.

In Ceylon under favourable conditions *Hevea* trees will grow in height at the rate of 6 to 10 feet per annum for the first three or four years after planting. In girth the increase may be about 3 to 4 inches per annum during the first few years. After this the rate may be slightly increased until the lateral branches have completely met, and then growth becomes slower once more. The greatest development takes place after the third year. Some of the oldest trees in Ceylon, at 35 years of age, had a girth of over 100 inches, and were about 80 feet in height. In Malaya the average growth of young trees is still more rapid, and it has been stated that the girth of four-year-old trees in the Federated Malay States is generally equal to that of five-year-old trees in Ceylon. The age of the trees is always reckoned from the time of planting, and not from the date of sowing the seed.

Weeding.

We next come to the vexed question of weeding. It is the universal opinion of practical planters that clean weeding from the burn off is cheapest in the long run, and leads to better growth of the trees than any other system. Cover plants should not be regarded as a substitute for weeding. Leguminous plants if grown as a source of nitrogen are to be regarded as a part of the general cultivation and manuring of the estate. These processes are additional to weeding and should not be looked upon as an alternative. Weeding is almost universally done by hand, but on suitable land the use of agricultural machinery is strongly to be recommended during the earlier stages of growth. The disc harrow drawn by oxen is a most effective implement for keeping down weeds and preserving a surface mulch, whilst on large estates steam power might very well be introduced. However, the presence of drains or unevenness of the land renders the use of such implements impossible on a large majority of rubber estates.

In connection with weeds and the use of cover crops, one remarkable fallacy, which has been again and again repeated, requires to be very clearly pointed out. It has been asserted that a close cover of leafy plants is a preservative against the effects of drought; and the moist surface of the soil beneath such a crop has been pointed out as evidence of the conservation of moisture. It is perhaps difficult at first sight to realise that a dry powdery surface is losing less moisture than a moist surface, but a little thought will show that this is certainly the case. The best preservative against the effects of drought is a thick covering of dead leaves, such as is actually present under old rubber when the leaves have fallen, an event which occurs towards the beginning of the dry season in Ceylon. Failing such a mulch, the next best form of protection is afforded by keeping the surface of the soil loose and powdery, in the form of what is known as a dust mulch, since a caked surface evaporates much more water than a loose surface. A leafy crop however is capable of evaporating three or four times as much water as a bare surface of soil. In times of drought therefore a thick covering of living weeds is a special danger, since the weeds draw off the supplies of water available for the rubber roots, and the crop is liable to suffer severely in consequence.

On steep slopes a cover crop may be of distinct utility in checking soil wash, especially if it is grown in definite lines across the slope, and periodically cut down and laid along the contours in order to form a series of miniature terraces. For this purpose various species of Crotalaria, Indigofera and Tephrosia may be used with advantage, since these are leguminous plants which also collect nitrogen from the air.

Intercrops.

If wide planting be adopted, some kind of intercrop seems to be desirable in order that some revenue may be obtained during the period before the rubber comes into full bearing. In Ceylon, tea is commonly used for this purpose, or more frequently the rubber is planted at wide intervals through existing fields of tea. In this case the growth of the rubber must be expected to be slow. Rubber has also been planted somewhat widely through cacao, but this practice is objectionable owing to the identity of several diseases to which the two crops are subject, notably canker. Cotton has been suggested, but the climates required by the two crops are widely different, since a prolonged dry season is necessary for the successful harvesting of the fibre. In Sumatra and Java coffee is frequently grown in conjunction with the early stages of rubber, and many other crops have been suggested. Amongst these, Indigo is one which seems to deserve more attention than it has hitherto received. Indigo is an efficient gatherer of nitrogen, and in the case of this crop practically the whole of the materials removed from the soil, together with the additional nitrogen, can be returned in the so-called *seet* in a readily available form. Indigo cultivation however requires special knowledge and appliances, and these it might not be worth while to introduce, owing to the comparative shortness of the period available before the shade of the rubber prevents the further growth of such a crop.

Schidrowitz and others consider that no intercrop has yet been proposed which will give a return greater than the relative loss occasioned by the slower growth of the rubber, so long as the price of rubber remains high. The question is not however finally settled, and much depends upon whether it is decided to plant the rubber trees at the distance at which they will finally stand, or to adopt close planting and subsequent thinning of the plantation.

Plate VI

Hevea Rubber and Tea

If the growing of an intercrop is decided upon, care must be taken to allow plenty of space for the growth of the rubber in its early stages. As the trees develop, the shade of the rubber will generally bring the life of the subsidiary crop to a close after it has served its purpose. All traces of the latter must then be cleared away, and no dead remains be left to serve as a centre for the spread of fungus diseases.

Cultivation and Manuring.

Few data exist with regard to the effect of cultivation on growth. Frequent deep forking has been tried on a small scale in Ceylon, and has apparently no harmful effect upon the trees. It is hardly to be expected that this expensive operation can lead to much financial profit when carried out over wide areas, although on sloping ground an occasional forking has a marked effect in checking wash. Mr Tisdall informs me that cases of wonderful development in growth have occurred as the result of annual deep forking in fields which were thought unsuitable for *Hevea*, owing to the slowness of growth during the early stages. Where the conditions are favourable for the employment of agricultural machinery, deep cultivation may profitably be employed between the rows of rubber, in connection with the eradication of weeds during the early stages of growth. More often than not the use of such machinery will be prevented by the frequency of drains, by the presence of rocks or other obstructions, or by excessive gradients. Later on it is probably better to conserve the mulch of fallen leaves and to imitate forest conditions as closely as possible.

With regard to the use of manures, no really reliable experimental work has yet been published, and opinions can only be based on general principles and on plantation experience. The latter shows considerable unanimity in favour of the use of artificial manures on the poorer classes of soils. Such manures, especially those containing nitrogen and potash, are said to exert a markedly beneficial effect on growth, yield and renewal during the earlier stages of the life of the tree. The application of potash is said to be specially beneficial in connection with renewal of bark. Phosphorus is found in practice to have less influence on the functions named ; and this is what would be expected on general agricultural principles, since the use of phosphates is generally closely associated with the formation of seed. The application of phosphates may therefore perhaps be suggested in connection with seed-bearers. Manures containing potash have the advantage of being consider-ably cheaper than nitrogenous and phosphatic manures. An excess of nitrogen is to be avoided, as it is said to make the trees top-heavy and brittle. Such an excess is well known to lead to an extravagant production of foliage in the majority of plants. Apart from this, it does not seem likely that the application of artificials in moderate quantities can have any deleterious effect upon a rubber plantation. The experience of the best authorities in Ceylon shows that the annual application of well chosen artificial manures leads to increased growth and yield and to a better renewal of the bark as well as to some immunity from disease.

Green Manuring.

Nitrogen is the most expensive of all the substances required by growing plants. A sufficient supply of nitrogen can most cheaply be added to soils, which are deficient in this respect, by the practice of green manuring. This method consists in growing some plant which is capable of taking up nitrogen from the air. The nitrogen is added to the soil by cutting down the plants at the proper stage of growth and either burying them or using them as a mulch. It is important therefore that this process should only be undertaken when there is plenty of labour available for dealing with the green crop at the proper stage. The plants used for green manuring belong to the natural order *leguminosae.* They may conveniently be divided into (1) creepers and climbers, (2) low growing bushy plants and (3) trees.

Creepers and climbers are not generally to be recommended. They are difficult to deal with, are liable to smother the young rubber trees if neglected, and form a convenient cover for snakes and other vermin.

Among low growing bushy plants some of the best, according to experiments carried out at Peradeniya, are

Crotalaria striata, Indigofera arrecta, Tephrosia candida
and *Leucaena glauca*. The last named plant grows into
a good sized shrub if permitted to do so, but it can be
kept coppiced and then provides a heavy mulch of
green shoots. Any of the above named plants may be
sown broadcast on level land or in transverse lines
on sloping ground. A maximum of green material is
generally available just before flowering, after four or
five months growth, when the plants have reached a
height of from four to six feet. The plants are then
cut down close to the ground and mulched round the
trees. On slopes they may be spread in the form of a
crescent some feet below each tree, in order to lead
to the formation of small natural terraces. Such a
mulch is specially valuable in times of drought, but
the method can only be applied with safety in countries
where the drought begins at stated periods. The cutting
and mulching must be carried out at the beginning of
such a period, otherwise the green crop will dry up
the soil by evaporating water. The dry stems left after
cutting lead to no appreciable loss of water, whilst the
green mulch is a valuable protection. After cutting,
the plants throw up fresh shoots, and in a moist
climate they can be cut repeatedly at intervals of four
or five months. Ultimately their growth is checked
by the shade of the rubber and they should then be
uprooted.

In situations where it will grow freely, one of the
best types of tree for green manuring is the dadap,

Erythrina lithosperma. This may be planted alternately
in the rows of rubber, and can be conveniently raised
from cuttings after a stock has been obtained from seed.
The best cuttings are stout branches five or six feet
long and three or four inches in diameter. These are
planted like so many posts, and develop into good
sized trees in little more than a year. Care must therefore be taken that their growth is not allowed to interfere
with that of the rubber. The green branches are
pollarded at suitable intervals and mulched as already
described in the case of smaller plants. The shelter
afforded by the dadaps is often valuable while the
rubber is young, and their presence tends to draw up
the latter into tall saplings and to check branching
close to the ground.

Shade and Wind Belts.

Rubber requires no permanent shade, and indeed few
trees would be tall enough to afford it. The protection
of dadaps, as already described, is often valuable in the
early stages of growth, and this is specially the case in
wind-swept situations. The interplanted dadaps afford
an excellent protection from the wind, but they should
not be allowed to top the rubber for too long a period.
Dadaps or similar trees may also be grown conveniently in belts across the direction of the prevailing
winds in cases where interplanting is considered
undesirable. They may also be introduced where a

low growing intercrop has been planted which affords no protection.

Pruning.

Except for the removal of dead branches, pruning is not generally recommended in the case of well grown rubber trees. Weak drooping branches which prevent the access of sunlight to the tapping area may, however, often be removed with advantage. Formerly a process known as thumb-nail pruning was advocated, in which the terminal bud of the tree was nipped off when a height of 12 to 15 feet had been reached. The result of this treatment was to cause a forking of the tree into two or three main branches, and it was claimed that the rate of increase of the girth of the main trunk was thus accelerated. The objection to the method lies in the fact that the fork becomes a point of weakness, and the tree becomes very liable to split at this point if at all exposed to wind. Such a fork also forms a convenient point of lodgement for fungus spores, and may thus lead to the origin of disease. The method is specially to be avoided in countries where Pink disease is prevalent. One form of pruning is, however, often necessary in the case of young trees, especially when widely planted. Any lateral branches which occur below a height of about ten feet should be cut off flush with the stem as early as possible, in order that the future tapping area may not be obstructed.

Thinning out.

There can be no doubt that thinning out is a problem which will soon have to be faced on the majority of estates, and on many the process is already in operation. The most important principle to be observed in thinning is to disregard the position of the trees. In the opinion of the writer, it would be a great mistake simply to remove alternate trees regardless of their individual value. Under ordinary conditions, whatever the position of the remaining trees may be, and however large the gaps caused by felling, the roots will soon explore the whole of the remaining soil, and the branches will come after a time to occupy the remaining air space. In the first place then, all weakly or diseased trees should be removed, and secondly all those which are found to give the poorest yields of rubber. Large trees may have to be taken down in sections, since their fall would lead to injury of the surrounding trees. The stumps must next be dug out or extracted. The site of the old tree should be thoroughly dug over and well limed, and as much of the roots as possible taken away. All dead wood and roots must be burned. The older parts of *Hevea* make a fairly good fuel which may be utilized in the factory.

Oil from Rubber Seeds.

It has been suggested that the collection of the seeds which are now produced in large quantities on *Hevea* plantations may form the basis of a profitable minor industry, since the kernels contain about 40 per cent. of an oil which is similar to linseed oil, and has been favourably reported upon by chemists at home. The suggestion to lease out the right of collecting the seeds does not appear a very happy one, owing to the opportunities for thefts of rubber which such an arrangement would afford. But at times of abundance the collection could be readily carried out on the estate by children too small to take part in the tapping and other regular work.

Hevea seeds weigh about 7 lb. per thousand, and the cost of collection last year was given by the Superintendent of the Peradeniya Experiment Station as under half a cent a pound. As the kernels represent approximately half the weight of the seeds, a ton of kernels will not cost more than Rs. 22.50 to collect under these circumstances. To this must be added the cost of decorticating the seeds, and for this purpose machines are under trial, with which it is hoped to perform the operation at a comparatively small expense. It is possible that the development of this subsidiary industry will form an appreciable addition to the value of rubber estates in the future.

Labour.

In densely inhabited countries like Java and South India, where the native population is obliged to work hard in order to obtain a meagre living, efficient labourers for employment on plantations are readily available at low wages. These two countries also provide a large number of emigrants to other parts of the Eastern Tropics which are less favoured in this respect, owing to the indolence or small numbers of the native population. Even in India a good deal depends upon the position and healthiness of the plantation. In every country some estates are always well supplied with labour, whilst others are exposed to constant difficulties in this respect.

Ceylon has also a considerable native population, and the number of Sinhalese labourers on rubber estates is increasing. The supply is however by no means equal to the demand, and the deficit is made up exclusively by immigrant Tamils and kindred races from the neighbouring peninsula. In Sumatra the labour employed on the rubber estates is principally immigrant Javanese, emigration being permitted by the Government of the Dutch colonies subject to certain regulations. The Malay Peninsula is comparatively thinly populated, and the Malays do not take kindly to regular work on plantations. There is also a considerable Chinese population, of which many are employed in the mining industry ; but Chinese labour has not been found entirely satisfactory for harvesting rubber, although the Chinese are widely employed in clearing and planting. In addition to a certain number of Chinese, Javanese and Malays, the rubber estates are principally supplied with indentured labour from

Southern India. Indeed, the drain of coolies from favoured districts in South India has recently been so great that the local planters are finding a difficulty in supplying their own needs, and there is a tendency to place obstacles in the way of further emigration. Large areas still exist, however, which have not yet been exploited by the labour agencies, and a good deal can probably still be done in this direction.

The number of labourers employed on estates of all kinds in the Federated Malay States in 1910 was approximately as follows:—

TABLE XXVII

Labourers on Rubber Estates in Malaya in 1910.

Tamils	...	...	99,000
Chinese	...	...	46,000
Javanese	...	...	18,000
Malays	...	...	14,000
Others	...	...	2,000
		Total	179,000

The great majority of the above were at work on rubber estates.

The number of Indian Tamils in Ceylon is probably nearly half a million, of whom perhaps a quarter are employed in the rubber planting industry. Every year a considerable proportion of these labourers returns to India by way of Tuticorin. The Manaar Railway is expected to be open next year, when the Tuticorin route will be little used. This increased facility will probably increase the flow of immigration.

Where recourse is had to immigrant coolies, the cost of their introduction must be paid by the estate, and in addition some advance is usually made, which may or may not be wholly recoverable from the labourers' wages within a certain number of years. In recent years, owing to the competition for labour, the system of advances in Ceylon has given rise to much abuse, and a state of affairs has arisen which is entirely opposed to the best interests of employer and employed alike.

The rates of wages paid to immigrant coolies, although sufficient to attract large numbers to leave their native country, are not very high according to western ideas. A good worker on a Ceylon rubber estate may earn from 50 to 60 cents a day (8d. to 10d.) and in the Malay States rather more. The governments of these countries have recently introduced legislation to protect the interests and welfare of the labourers, and increased attention is paid to their comforts and sanitation, whilst hospitals and schools are now being provided in all districts.

Other Expenses.

In addition to the work already described, some of the other items of expenditure involved in opening up an estate may here be briefly enumerated. In the first place, there are the salaries of the Superintendent and his assistants, and the wages of native overseers, watchers and guards. Houses must be provided for all these in addition to "lines" for the coolies, stores, hospital and dispensary, cart sheds and other buildings. On large estates horses and stabling must be provided for the Superintendent.

A full survey of the estate is necessary, and a large scale map should be prepared showing watercourses and other physical features, together with the subdivision of the estate into fields of 50 or 100 acres. Carts or lorries or boats will be required, according to the method of transport adopted. The food of the coolies is mainly supplied by the estate and charged to their account. Other items are medicines, cattle food, fuel, etc.

Tools.

The large hoe universally employed for digging in the barefooted East is known as a chankol in Malaya and as a mamoti in Ceylon. Digging, draining and a certain amount of weeding are done with this implement. Felling axes, cross-cut saws and bill hooks are required for clearing, and for lighter work the parang is used in Malaya and the katty in Ceylon. For nursery work, rakes and watering cans are required, and a kind of crow-bar is generally used for digging holes for planting. These, with pruning knives and saws, complete the list of tools required on a rubber estate before the time for tapping arrives.

In addition to the above, the Superintendent will require measuring tapes and chains and a few simple surveying instruments.

I am indebted to Mr W. N. Tisdall for the following estimate for planting 500 acres of rubber and for bringing it into bearing under average conditions in Ceylon :

Purchase price of Crown Land may be any figure between Rs. 75 and 250.

First Year (opening).

	Per acre
Felling and clearing	15.00
Holing 20′×20′, 108 @ 6 cts.	6.48
Filling	1.00
Liming	1.50
Weeding, 6 months @ 2/-, 6 months @ 1/50 ...	21.00
Planting	1.50
Nurseries and plants	5.00
Fencing (Barbed and Mesh)	10.00
Roading, including blasting, etc.	12.50
Draining ,, ,, ...	17.50
Supervision	15.00
Buildings, 50 coolie rooms @ 25/2	2.50
Bungalows, etc.	3.00
Coolies, advances, 200 @ 60/-	20.00
Medical, etc.	5.00

Second Year.

General upkeep and supplying, etc.	50.00

Third Year.

General upkeep and supplying, etc.	45.00

Fourth Year.

General upkeep and supplying, etc.	40.00

Fifth Year.

General upkeep and supplying, etc.	35.00
Permanent Buildings, etc.	120.00
Rs.	426.98

The sixth year should show a good return. Under favourable circumstances this estimate could be reduced, but it provides for the best class of work.

CHAPTER VI

Introductory.

PREPARATIONS for tapping the trees are generally begun when the plantation has reached an age of four or five years from planting. In each field over a certain age—usually not less than four and a half years—the trees are measured and numbered. All trees over a certain girth—usually not less than 18 inches at three feet from the ground—are then marked out for tapping on one or other of the systems presently to be described. The age at which the trees can first be tapped is largely determined by the thickness of the available bark. Experience has shown that trees of 18 inches girth and upwards can generally be lightly tapped without injury. This size, as we have seen, corresponds to an age of from four to six years. At this age the bark has reached a thickness which is capable of sustaining the effects of careful tapping and of making a satisfactory renewal, whilst yielding a certain amount of latex. From trees of a smaller size than this the yield hardly repays the cost of tapping, and the risk of serious injury is great owing to the thinness of the bark. If the planter can afford to wait until the trees have reached a somewhat larger girth so much the better.

Before describing methods of tapping in detail, we may give a general account of the ordinary routine of an estate on which the harvesting of the rubber is in progress. A start is made in the very early morning, since the earlier the trees are tapped the more freely does the latex flow. Each tapper has a certain number of trees assigned to him or her. In some cases each worker is allotted the task of tapping a certain number of trees and is paid by the day. On other estates payment is made according to the quantity of rubber obtained. In such cases each tapper has the run of a somewhat larger area. In either case the quality of the work performed requires thorough supervision, since the least carelessness in tapping may lead to serious injury of the bark. At the first round the trees are tapped and collecting cups of tin, glass or coconut-shell are placed in position to receive the latex. When the tapping round is completed, the coolie makes a second round provided with a pail of enamelled iron, into which the contents of the several cups are poured. The latex is carried to the factory, where the quantity obtained by each tapper or group of tappers may be separately coagulated, and the wet rubber weighed, before it passes on into the general stock. A third round may then be made in order to scour the cups and to collect the scrap rubber which has congealed in the cups or upon the bark of the tree. More often this work is relegated to a less experienced hand. Periodically also the shavings of bark which have fallen upon the ground are collected, and even the earth upon which latex may have trickled is not wasted; from both these sources low grades of rubber are extracted in the factory.

Tapping processes may be divided into methods of incision and methods of excision. In the excision or paring methods, which are almost universally employed in practice on estates, a thin shaving of bark is removed from the tree at each tapping. Incision methods are designed to extract the latex by pricking or gashing without removal of bark.

Incision Methods of Tapping.

Methods of incision or pricking have many points to recommend them, at least in theory. If we are to credit the dictum of Mr Herbert Wright, that the best method of tapping is the one which leads to the greatest flow of latex with the least possible removal of bark, a pricking system in which no bark is removed should be superior to any method of paring. There are however many other points to be taken into consideration besides the flow of latex and the removal of the bark. It may therefore prove instructive to discuss the disadvantages under which some systems of pricking labour as compared with good paring. From such a discussion we may hope to arrive at an understanding of the features which should characterise an ideal system of pricking.

The method of incision was employed by Trimen in Ceylon prior to 1888, but the precise method adopted was not exactly specified. Among the earliest methods of incision practised upon planted trees, of which we have a satisfactory record, was the one employed by Parkin in experiments at Henaratgoda. Parkin's method, which was based on those used in the collection of wild rubber in Brazil, was to make V-shaped cuts in the bark by the aid of a hammer and chisel. With this method two definite defects are associated in practice. A separate collecting cup is required for each V, whilst the surface of the tree becomes rough and lumpy owing to the irregular renewal thus induced, until further tapping becomes a difficult matter. Moreover, except in the case of old trees with very thick bark, it is impossible to avoid more or less extensive injury to the cambium. The method is quite unsuitable for young trees.

The fact is that no method of large incisions truly conforms to Wright's definition of good tapping, since such incisions entail the destruction, if not the removal, of a considerable mass of living cortex. This destruction goes much deeper than that caused by careful paring, and often involves the cambium itself. On the other hand, a small clean prick, such as may be made

with the point of a pen-knife, generally heals up completely without any sloughing off of cortical tissue, in
spite of the fact that the blade has actually penetrated
to the cambium. Such a prick also leads to a considerable flow of latex in comparison with the size of
the wound inflicted.

The time required for making small pricks singly
is clearly prohibitive of their use as a practical method.
For the rapid infliction of a large number of pricks the
rotating spur-shaped pricker was introduced by Bowman
and Northway in 1905, and has enjoyed a considerable
vogue in Ceylon. A single stroke from one of these
tools produces a row of pricks running across the bark
of the tree. Owing to the difficulty of collecting the
latex thus liberated by any other method, the use of
the spur-shaped pricker was generally combined with
a shallow system of paring. The pricker was run
along the paring cut either on the same day or on the
day following, and the cut provided a channel down
which the latex could flow. The great theoretical
objection to this method lay in the closeness of the
pricks inflicted on successive days. Although the bark
of a healthy tree generally recovered from the operation,
there was a considerable sloughing off of cellular tissue.
Moreover, in the freshly pricked region the blockage of
phloem tubes necessarily produced must have amounted
to partial ringing of the tree. It seems clear too, that
in some cases the use of this method was associated
with an epidemic of woody nodules (see Chapter VIII).

Plate VII

Photo C. Northway

Hevea Tree pricked with Serrated Knife

Finally, there is clear evidence that a better yield, without any greater loss of bark, is obtained in the long run by simple paring than by combined paring and pricking.

Pricking Methods under Trial.

A method has been developed on an estate in the Southern Province of Ceylon which seems to avoid many of the drawbacks hitherto described. The trees are tapped on a herring-bone system, and each rib of the herring-bone is simply represented by four small pricks made by a single insertion of a " serrated knife." This pricker has square teeth a quarter of an inch wide, with a similar interval between each tooth. They are thus practically identical with what is known as Macadam's comb pricker, except that the number of teeth is only four. A shallow conducting channel is cut vertically in the bark of the tree, and with a little assistance from the tapper the latex flows into the channel and down the tree. The original incisions, covering an area 2 inches in breadth, are made at vertical intervals of a foot, and on each day following a similar set of incisions is made at $\frac{1}{2}$ to $\frac{3}{4}$ inch below the old ones. When the first area is completed after 24 days' tapping, the whole operation is repeated on the opposite side of the tree. Subsequently similar figures are intercalated between the old ones until only one or two narrow strips of bark are left untapped. A return is then made to the original area after a period of rest, which is more or less extended according to the freedom with which the latex is found to flow.

The method here described is open to the same abuse as all other pricking systems, in the temptation which it presents of overtapping small trees. But if no trees under 18 inches in circumference are tapped, and if the method is adopted of resting any field which shows a falling off in yield, the system seems to be largely free from theoretical disadvantages. According to Mr Tisdall, a noticeable feature in this method of pricking is the immunity of the tree from canker and bark-rot in comparison with the paring system, and the healthy appearance of the foliage after an extended period of tapping. So far the disadvantages appear to be the difficulty of collection in wet weather and the extra cost of tapping as compared to paring. On the estate where the system has been evolved, highly satisfactory yields are said to be obtained by this method. The system may be recommended for *trial* elsewhere, bearing in mind that a method of tapping which suits one district is often quite unsuccessful in another.

Another incision system, suggested by Mr Kelway Bamber, has been on trial for some time by the Botanic Department of Ceylon. The method is deserving of mention on account of the large yields which have been obtained from young trees without apparent damage.

Bamber's method consists in cutting two vertical

channels on opposite sides of the tree, and pricking down these from top to bottom with transverse cuts an inch apart. The cuts are made with a thin blade about three-quarters of an inch wide and rounded at the end. On alternate days two similar channels are cut an inch to the right of the former ones and pricked in the same manner. When the whole of the circumference has been tapped in this way, the tree is rested until the beginning of the second month from the commencement of tapping, and the process is then repeated, the same vertical channels being cut smooth and re-pricked. In this way the larger trees are tapped more frequently than the smaller ones. On a particular acre bearing 147 trees the average number of tappings was 75 in a year. The trees were just five years old at the commencement of tapping, and ranged in circumference from 10 to 21 inches. All the trees were tapped without exception, and the average yield for twelve months was 18 ounces per tree, or 164 lbs. per acre. This is a remarkably high yield for 5—6-year-old trees at an elevation of 1500 feet in Ceylon. At the end of the year only a few of the smallest trees appeared to have suffered any harm from the process.

The chief objections to this system are the large amount of labour required, and the large proportion of the rubber which is obtained in the form of scrap, namely, nearly 50 per cent. of the total quantity. It is however a method which may well be adopted in tapping trees which it is ultimately intended to thin out on closely planted estates. How far it can be adopted as a permanent system of tapping can only be determined by further trial.

Excision Methods of Tapping.

At the present time the method which is almost universally adopted on estates is paring on some system or other. The results are generally so satisfactory that paring is not likely to be given up in favour of any method of pricking, until very clear proof of the superiority of the latter is forthcoming.

The precise system of paring will partly depend upon the period which is to be allowed for bark renewal. At the present time it is generally recommended that the renewed bark should not be touched until a period of four years has elapsed from the beginning of tapping. There is some reason for believing that even this interval, which is longer than that formerly adopted on many estates, may be profitably increased. It is probable also that a longer period is required for each successive renewal.

As regards the size of tree upon which paring may safely be begun, it is generally considered that trees less than five years old from planting should not be tapped at any point where they are less than 18 inches in girth. The reason for this restriction is that where the circumference is smaller, the bark is so thin that it is almost impossible to pare without causing injury to the tree. Some authorities would increase the minimum size allowed to a girth of 20 inches or more.

If it is decided to allow four years for the renewal of the bark, the simplest system which can be adopted is to divide the circumference of the tree into four equal parts, and to tap these successively by the half-herring-bone method. Each quarter section of the tree will then represent a year's tapping. The quarter section opposite the first should be tapped second, and then the other two sections successively, in order to preserve the symmetry of the tree as far as possible.

Systems of Paring.

The system of tapping adopted in the experiments described in Chapter IV was one of superimposed V's. The cuts were made at an angle of 45 degrees with the horizon, and the vertical distance between successive cuts was one foot. In describing other systems we may assume for the present a similar angle of cut, and a similar distance between successive cuts where there are more than one. The half-herring-bone system (Fig. 16) referred to in the last paragraph consists of a vertical conducting channel into which a single series of slanting cuts leads on one side only. Such a system may be derived from the superimposed V system by obliterating one limb of each V. Reasons have already been given for confining such a system to one quarter of the

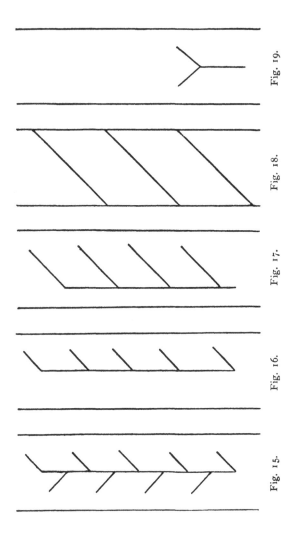

Fig. 19.

Fig. 18.

Fig. 17.

Fig. 16.

Fig. 15.

circumference. A similar system, extending over half the circumference of the tree, has however been widely used on estates, and is known as the half spiral system (Fig. 17). The full spiral system was once popular but is now almost obsolete, as it proved too drastic. In this system one or more similar cuts extended right round the tree, or even girdled it more than once. There was no vertical channel, but each spiral cut drained into a separate collecting cup. In the full herring-bone system (Fig. 15) there is a conducting channel with slanting cuts on either side. The lateral cuts enter the central channel alternately instead of meeting as in the V system. The herring-bone method is generally preferred to the V system because in the latter the bark at the tips of the V's is said to be specially liable to injury. The full herring-bone system is often used on large trees, on which the lateral cuts would have to be of considerable length if the half-herring-bone system were employed. To mention only one other system—on young trees in Malaya a single basal V (Fig. 19) is often tapped. By this method trees can be tapped close to the ground, although the bark at a higher level is not yet ready for tapping. The basal system is also being used on older trees on some estates, and it is said that the yield is almost as good as when two or three cuts are employed, whilst there is of course a great saving of bark.

Marking the Tree.

When the age for beginning tapping has been reached, all the trees on a particular field are generally numbered consecutively. The trees are measured, and those which exceed a certain girth are marked out for tapping. It may be useful to describe a method of marking in some detail, in order that the system adopted may be clearly understood. The method of tapping to be described is the half-herring-bone on a quarter of the tree.

We will suppose a tree to be 20 inches in girth at 4 feet 6 inches from the ground. At this level a horizontal line is stencilled round the tree. From two points 5 inches apart on this line two vertical lines AB, CD are ruled down to the foot of the tree by the help of a straight edge. The tree may be perhaps 24 inches in girth a foot from the ground. If so the lines will be 6 inches apart at the point S. These lines should be so arranged that the equidistant points B, D, F, and H fall between the main lateral roots of the tree, which generally project above the level of the ground; otherwise there may be a difficulty in getting the collecting cups to stand upright when tapping is in progress.

Next a mark is made at P, 5 inches below A, so that AP is equal to AC. A line drawn from P to C will then make an angle of 45 degrees with the horizontal. Marks Q, R, S, T, U, and V are made at intervals of a foot below P and C respectively, and lines are drawn joining up *PC, QT, RU*, and *SV*. These lines represent the positions of the original tapping cuts. The lower lines will make angles of rather less than 45 degrees with the horizon, but the slope will not be so much less as to make any serious difference in the result of tapping.

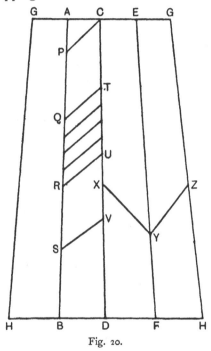

Fig. 20.

This method of marking will be found more accurate than that of measuring up from the ground, since the ground level on different sides of the tree is liable to vary considerably. The lines are best marked boldly with a tapping knife, in order to leave a permanent trace on the surface of the trunk.

From P nearly to B a broad shallow groove is cut with the tapping knife in order to form the conducting channel for the latex. The line CD should also be marked out with a clear cut, in order that the tapper may keep within the limits of the area allotted to him. It will further conduce to accurate work if additional guide lines are stencilled between the original tapping cuts, and parallel to these, at intervals of about 3 inches, as shown within the area $QTUR$.

In the above example the original tapping cuts have been placed 12 inches apart, in order to allow of a year's tapping on alternate days at the rate of 15 cuts to the inch—180 cuts in all. It is reported that in many parts of the Malay Peninsula the average width of the daily shaving removed is considerably less than one-fifteenth of an inch. When this is the case the original cuts may be placed somewhat closer together. On the other hand, if daily tapping be adopted it may be necessary to place the original cuts further apart.

In the case of smaller trees—those which are less than 20 inches in circumference at a foot from the ground—tapping may be begun on a single basal **V**, as shown at XYZ in Fig. 20. After this **V** has been tapped for a year, the tree will probably be large enough to allow of placing two or more half-herring-bone traces on the opposite side of the tree. In this case the areas $ACBD$ and $AGHB$ would next be tapped in succession, followed by a further half-herring-bone on the area $CXVE$ above one-half of the original **V**.

The Paring Process.

The actual process of tapping is as follows. The opening cuts are narrow grooves cut in the outer bark and extending to within about one-eighth of an inch of the cambium. It is important that the depth of the groove should be exactly the same throughout its length. The distance from the cambium can be gauged by pricking with the blade of a pen-knife. Further tapping consists in removing a thin shaving from the lower side of the original groove on each occasion, always to exactly the same depth. Coolies soon become very expert at this work, but constant supervision is necessary in order that no damage may be done. If the tapper makes a mistake and cuts too deeply, injuring the cambium, he should be instructed to make three or four very shallow cuts immediately below the wound, so as to leave a thick piece of bark from which healing can take place. If tapping is continued to the normal depth—no matter how carefully it is conducted—the wound is almost certain to spread. In the best tapping the cuts are made to slope slightly inwards in order to form a suitable groove by which the latex can reach the conducting channel. The latex may need a little encouragement before it will flow smoothly in the right direction, without overflowing the channel. The merits of good tapping consist in taking off the thinnest possible shavings compatible with a free flow of latex and preserving an even depth of cut, close to the cambium but never touching it. The cuts should be as clean and as sharp as possible.

Tapping Tools.

Of tools used in paring there is a great variety upon the market. Many of these are highly effective, and it is not our business here to puff the goods of any particular maker. The writer's view is that the simplest tools are the best, and that a great deal is to be said for the original and primitive forms—the gouge and farrier's knife—or slight modifications of these. Such tools are

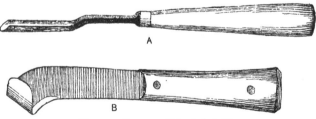

Fig. 21. Gouge and Farrier's knife.

very widely used in the Malayan plantations, but the Ceylon planter seems generally to prefer a more complicated weapon. The simpler tools may necessitate greater care at first on the part of the tapper, but once the requisite skill is attained, the greater adaptability of the simpler tool and the greater ease with which it can be kept sharp, enable the worker to preserve a very straight line and to remove a very thin shaving, keeping at the same time an accurate depth of cut.

Angle of Cut.

The fact that an angle of 45 degrees with the horizon is often adopted for the lateral cuts is largely a matter of convention. A more acute angle than 45 degrees is seldom used, but flatter cuts are adopted on a good many estates. Careful and prolonged experiments are required before any definite statement can be made as to what is the best angle under various circumstances and with different systems of tapping. A flatter cut opens the same number of latex vessels as a more acute cut extending over the same horizontal distance, and consequently releases an equal amount of latex with the removal of less bark. On the other hand, the latex flows more freely along a steeper cut, and a smaller proportion of scrap is thus obtained. When the consistency of the latex is moderately thin, the angle of the cut can probably be reduced to one of about 30 degrees without any loss of flow and with some slight saving of bark.

Direction of Cut.

In the half-herring-bone and half-spiral methods of tapping, the cuts are usually made from the right to left of the operator. There is a considerable body of evidence to show that more latex is obtained in this way than from cuts made in the reverse direction. The reason for this phenomenon is not entirely clear. It may partly be associated with the fact that tapping from right to left is easier for a right-handed person, and is therefore carried out more efficiently. But this does not seem sufficient to account for the whole of the difference.

Distance Between Successive Cuts and Yield at Different Levels.

In various systems of tapping, a distance of one foot between successive cuts has been very widely adopted. Here, again, prolonged and laborious experiments are required in order to ascertain what is the most profitable distance to adopt in the case of each system and under various conditions. The evidence contained in the following paragraphs, which bears partly on this question, requires to be greatly elaborated.

When a fairly old tree is first tapped by a herring-bone or multiple V system, adopting the usual interval of a foot between the cuts, it is found that the latex flows in nearly equal quantity from each cut of the vertical series. As the bark between the several cuts becomes used up, a relatively larger proportion of latex is found to flow from the lowest cut; and some of the upper cuts may even cease altogether to yield latex, whilst there is still an inch or more of bark remaining untapped. It is probably best in all cases to leave the last inch or so of bark untapped, and to continue tapping the lowest cut only, if the time has not arrived for passing on to a fresh area of bark. When the same section of the tree comes to be tapped a second time, it may be marked out an inch higher than before, in order that the unused slips of bark may now be tapped first.

In an experiment at Henaratgoda 29 old trees were tapped daily on six wide V's, spaced at vertical intervals of a foot. During the first month's tapping the three upper cuts yielded together a daily average of 558 c.c. of latex (nearly a pint), and the three lower cuts a daily average of 561 c.c. The quantities were therefore almost exactly equal, although the latex from the lower cuts contained a slightly higher proportion of rubber. During the six months' tapping, at the end of which the bark between the cuts had been almost entirely removed, the average daily yield from each of the six cuts was as follows:

TABLE XXVIII

Cut	1	2	3	4	5	6
Yield c.c. ...	77	86	98	82	83	400

Thus, when the bark between the cuts was nearly exhausted, the lowest cut was yielding as much latex as all the other five put together. In addition, the latex from the lowest cut was more concentrated and contained a higher percentage of rubber.

The trees described above were old, and possessed a good thickness of bark. In the case of younger trees we should probably expect the difference to be still greater. On the other hand, the rate of tapping was exceedingly rapid, the whole of the bark being removed from one side of the tree in six months. The evidence on the whole, however, seems to point to the conclusion that the distance between successive cuts may profitably be made greater than 12 inches. In other words, a reduction in the number of cuts on a given area of bark, as compared with the systems now generally adopted, would probably lead to more satisfactory results.

The above experiment shows further that in the case of mature trees there is little difference between the initial yields of latex from areas of bark situated at a height of 1—3 feet and 4—6 feet from the ground respectively. Generally speaking, there is a slight falling off in yield as we pass up the trunk from ground level, owing to the diminished thickness of bark and the smaller girth of the tree. In the lower part of the trunk of mature trees this falling off is not very rapid, and the fact that tapping is often confined to the lowest six feet of the trunk is largely a matter of convenience.

It must be remembered that the lowest cut of all drains a larger area of bark than any of the other cuts, a fact which partly accounts for the very much larger yield obtained from it.

Tapping Intervals.

The effect of different intervals between successive tappings on old trees has been discussed at some length in Chapter IV. So long as the yield from a young plantation continues to increase steadily, we have not at present sufficient grounds for recommending any change from the interval of one or two days which is usually adopted ; in fact there is evidence that in the case of *young trees* the yield per tapping may decrease as the interval between successive tappings is extended. In the case of older plantations of closely planted trees, if the yield shows a tendency to remain stationary or to fall off, we would suggest the trial of twice as long an interval without any other alteration of the system in use. We should not be greatly surprised if the planter were to find after some months that he is harvesting as much rubber as before, with an expenditure of only half the labour.

Difficulties and Precautions.

The physiology of latex production was discussed in Chapters III and IV as fully as is justified by our limited knowledge of the subject. The present section contains a few general remarks in relation to tapping. One of the chief troubles of the latex collector consists in the washing away of the latex which occurs in wet weather, owing to the rain water which streams down the trunk of the tree. Rainy days generally make up a large proportion of the tapping season, and if tapping is stopped whenever there is heavy rain, the loss of yield will be considerable. A simple system of guttering has recently been introduced in Ceylon, which can be fitted to the tree above the tapping area, with the object of keeping the latter dry.

Another reason for keeping dry the tapping area, lies in the fact that this region is more susceptible than any other to the attacks of disease germs. A constant watch should therefore be kept on the tapping cuts, and at the first sign of disease tapping should be suspended. There are two reasons for ceasing to tap a diseased tree; firstly in order to allow the individual tree to recuperate, and secondly to avoid the transference of infection to other trees by means of the tapping knife.

In the absence of visible signs of disease, certain cuts are sometimes found to yield little or no latex, whilst there is still a comparatively free flow from other cuts. When this is the case, the hint should be taken and the tree rested, otherwise definite disease is not unlikely to supervene. For a description of particular diseases and their treatment, reference may be made to Chapter VIII.

At the bottom of the conducting channel a tin spout should be fixed into the bark, in order to throw the latex clear of the surface of the tree into a cup placed on the ground below. The practice of pushing the edge of a metal collecting cup into the bark at each day's tapping may lead to injury, and is not to be encouraged.

In all operations dealing with latex the utmost attention must be paid to the virtue of cleanliness. Latex is as easily contaminated as milk, a substance with which it has many properties in common. Not only should the standard of cleanliness in the factory be at least equal to that adopted in an up-to-date dairy, but the avoidance of all kinds of dirt should begin from the moment when the latex first makes its appearance upon the surface of the tree. Collecting cups, paring knives, and the bark itself should all be kept scrupulously clean. All scrap should be carefully removed from the tree before paring commences. In order to save the precious hours of the early morning, this may very well be done during the afternoon of the day before tapping.

Plantation Yields.

Finally, some idea may be given of the actual yields of rubber obtained upon estates. The amounts recorded vary very largely on different properties, but the following figures may be taken as representing approximately the average yields obtained with moderate systems of tapping in Ceylon and the Malay Peninsula respectively.

TABLE XXIX

Age of trees, years	Ceylon		Malaya	
	Per acre, lbs.	Per tree, lbs.	Per acre, lbs.	Per tree, lbs.
4— 5	50	·4	100	·8
5—· 6	100	·6	150	1·0
6— 7	150	·8	200	1·3
7— 8	200	1·2	250	1·6
8— 9	250	1·5	350	2·2
9—10	300	1·9	400	2·5

Very much higher yields than these have been obtained under favourable conditions, and larger average yields may probably be anticipated in later years, now that the importance of careful and moderate tapping during the early stages is beginning to be fully realised.

CHAPTER VII

FACTORY WORK ON THE ESTATE

General.

THE latex, brought to the factory in tanks or buckets
as it is collected from the trees, contains as a rule from
30 to 40 per cent. of pure caoutchouc. The latter is
obtained in a state of considerable purity by the
successive processes of coagulation, washing and drying.
The method of preparing biscuit rubber by hand, as
practised during the earliest beginnings of the industry,
is now of little more than historical interest. It may
serve however to illustrate in their simplest form the
nature of the chief processes concerned.

In this method the latex is first strained through
a brass or copper wire sieve of small gauge, and then
generally diluted with one or more times its own bulk of
water. The diluted latex is next agitated with a dilute
solution of acetic acid, containing about one volume of
pure acid for every 1000 volumes of pure latex. The
acidified latex is then poured into round shallow pans to
set. Coagulation begins almost immediately, but the

pans are allowed to stand for several hours until a firm cake of rubber of the consistency of cream cheese has formed upon the surface. The remaining liquid is by this time perfectly clear and free from rubber. The cake of rubber is then removed, washed in clean water, and rolled out thin on a board with a common rolling pin or a bottle, in order to express as much water as possible. The resulting biscuits are then spread out to dry on racks in an airy and darkened room. In favourable weather the drying may be complete in from 15 to 20 days, if the biscuits do not exceed one-tenth of an inch in thickness. The resulting product consists of thin round sheets of rubber. If all goes well, the biscuits are semi-transparent and of a uniform honey colour.

The first step towards more elaborate processes of manufacture was the substitution of a simple rolling machine for the bottle or rolling pin; but with the increase of crops from large estates, hand labour has largely been ousted by the introduction of heavy machinery.

The Factory.

The site for the factory should combine a central position on the estate with accessibility to the main road or other means of transit leading to the port of shipment. A plentiful supply of good clean water is essential for washing. It is generally best for the factory to lie at the lowest possible level, since the incoming latex is heavier for transit than the outgoing rubber. Water power should be made use of if available, but this is comparatively rare on rubber estates. The source of power usually employed consists of steam or suction gas or liquid fuel engines, of which the latter are probably the more popular. A supply of steam may however be required for heating purposes.

The amount of power required depends upon the size of the estate. On a plantation of 500 acres provision will have to be made for turning out 1000 pounds of dry rubber daily at certain seasons of the year. The treatment of the latex cannot be delayed, and machinery must be available for dealing with the maximum crop at any season. For this output about 45 horse power will be required for the washing and crêping machines, in addition to about 8 horse power for vacuum driers, if these are adopted.

Several firms of engineers now specialize in the building of rubber factories and in the provision of suitable machinery. The details of structure and equipment best adapted to the needs of any particular estate can best be settled in consultation with a reliable firm. One of the most important details to be considered relates to the lighting of the factory. Plenty of light is desirable for working the washing, rolling and crêping machines, but in the subsequent stages of the process of manufacture an excess of light is to be avoided, owing to the injurious effect of light upon the rubber. Rooms intended for the drying and storage of the rubber require to be specially well screened from light, either by curtains or by painting the glass of the windows.

It is usual to begin with a factory of moderate size, which is capable of expansion as the crops of rubber increase ; and it should be borne in mind when drawing up the original plans, that provision may ultimately have to be made for dealing with a crop of more than two pounds of rubber daily from every acre of the planted area. Another essential point is the necessity for the utmost cleanliness at every stage of manufacture. Any carelessness in this respect is liable to lead to the production of discoloured or even tacky rubber, with a corresponding diminution in the value of the product. It is worth while to copy from modern hospital buildings the method of rounding off all corners where walls and floors meet. Internal surfaces, wherever possible, should be of smooth cement. Floors should be sloped so as to drain into cemented channels for greater ease in washing. A plentiful supply of clean water is essential.

Transport of Latex.

The latex is generally brought in from the field by coolies in enamelled iron buckets. On very large estates wheeled tanks are sometimes employed, and in some cases a system of small tram lines or monorails has been resorted to. Where the bulk of latex to be dealt with is very great, coagulation is often carried out in sheds in the rubber fields, and the wet rubber thus obtained is transported to the central factory for further treatment.

Coagulation.

The principal medium employed for effecting coagulation is acetic acid. Other coagulants are sometimes employed. The merits of hydrofluoric acid have been widely advertised, but the powerfully corrosive effect of this substance is a disadvantage. We do not know of any exhaustive comparison between the effects of hydrofluoric acid and those of acetic acid. Mixtures containing tartar emetic, formaldehyde and other substances have also been recommended.

Fresh latex is slightly alkaline. The chemical and physical nature of the process of coagulation is not altogether understood, but it is generally supposed that the phenomenon consists primarily in the precipitation of the proteids of the latex which are soluble in an alkaline medium. The coagulated proteids form a network in the meshes of which the globules of rubber are entangled, and the whole then contracts into a clot. An excess of acid leads to renewed solution of the proteids, so that either too much or too little acid produces incomplete coagulation. The special merit of acetic acid lies in the wide range of proportions in which it can be added to the latex, whilst still ensuring complete coagulation. Thus Parkin states that the acid can be added either in quantities four times below the proper amount or nine times above it, with very little waste of

rubber. Acetic acid has also less destructive effect on
the finished rubber than most of the mineral acids. It
is generally recognised however that an excess even of
acetic acid is harmful to the manufactured product. It
is therefore important that only just sufficient acid
should be added to ensure complete coagulation. The
exact amount required appears to vary considerably
under different circumstances. Parkin found that complete coagulation was produced by the addition of
·09 per cent. of pure acetic acid, or one part in about
1100 of pure latex. Working with latex from the old
Henaratgoda trees, we have found that this amount is
often insufficient, whereas on many estates a smaller
proportion is generally found to be effective. It is to be
recommended that in large factories the bulked latex
should frequently be tested on a small scale in glass
vessels, in order to ascertain the smallest amount
necessary. If the latex has been kept for some time,
its alkalinity may be reduced by the action of putre-
factive bacteria, and a smaller amount of acid will then
be required.

The correct amount of acid, well diluted with water,
is added to the latex, which is then thoroughly stirred
and allowed to stand. When crêpe is being manufactured the coagulation usually takes place in enamelled
buckets. From these the spongy rubber is removed after
an interval of half an hour or less and transferred at
once to the washing machines. For the manufacture of
sheet rubber the latex must be set in shallow pans. In

RUBBER PLANTING 159

these it is allowed to stand for some hours until a firm clot is formed, which can be lifted out in one piece and rolled into sheet. The pans vary in depth from 2 to 4 inches, according to the thickness of the sheet required, and a common size is 9 by 18 inches.

As opposed to coagulation, a centrifugal method of separating the rubber globules from the latex was suggested some years ago by Biffen. An electrolytic method of separation has recently been patented by Cockerill. In addition to these methods, various forms of mechanical churns and separators have been recommended for use in connection with acid coagulation. The ordinary method has a considerable advantage over all the last-named processes in the matter of simplicity, and it seems likely for the present to hold its ground. The centrifugal method has however a special use in dealing with *Castilloa*. At present the method of crêping is decidedly the most convenient and the most rapid for dealing with large quantities of latex.

Washing.

The majority of commercial rubber, as purchased by the manufacturer, contains numerous impurities in various proportions. The first step in dealing with all wild rubbers is therefore to subject them to a thorough process of washing and purification. The following mean values of the loss in weight on washing are given by Weber for different commercial rubbers.

TABLE XXX.

Loss of weight of rubber samples on washing.

Trade name	Loss, per cent.
Para, hard cure	15
Congo Ball	28
Ceara (Manicoba)	32
Borneo	48

One of the chief merits of plantation rubber from the point of view of the manufacturer lies in its high degree of purity, and the very small loss which consequently results when it is subjected to the washing process. In fact, in the near future, manufacturers who deal exclusively with plantation rubber will probably be able to omit the preliminary washing process altogether.

Owing to the clean character of the latex brought in to the factory on an estate, mere coagulation and drying would lead to the production of a rubber of a higher degree of purity than any wild kind. Washing is necessary however in order to remove the residue of the acid left over after coagulation, together with the soluble constituents of the latex. The latter serve no useful purpose, and are liable if retained to act as nutriment for moulds and bacteria.

The washing machines employed on estates are generally similar to those used by rubber manufacturers, but are constructed on a somewhat smaller scale. Such machines consist essentially of a pair of heavy steel rollers which revolve in opposite directions. The hand roller illustrated shows all the essential features of heavier machines which are driven by belts or shafting. The distance between the rollers is adjustable by special screws. A spray of water plays over the rollers, and a strainer is placed below the machine in order to collect any fragments of rubber which may be torn off in the

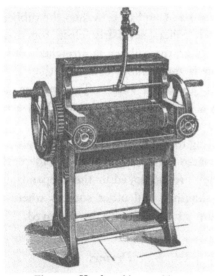

Fig. 22. Hand washing machine.

washing process. Three types of machine, differing somewhat in detail, are employed for different purposes, namely sheeting, crêping and macerating respectively.

Sheeting machines have smooth rollers which revolve at an equal rate of speed. The rubber sheets are simply passed two or three times between the rollers under a stream of water. The washing of sheet rubber is there-fore largely superficial.

Crêping machines have grooved rollers which revolve at different rates. The grooves may be arranged in a diamond pattern, or the roller may have plain parallel grooves disposed either longitudinally or in a slight spiral. The relative rates of revolution of the rollers varies between 2 : 1 and 6 : 5. When the rubber is passed between such rollers, set fairly close together, under a stream of water, the rubber is stretched and torn and the washing is much more complete than if the rollers revolved at an equal rate. The rubber leaves the machine in a long lace-like strip, which is a convenient form for rapid drying.

Macerating machines are similar to crêping machines, but have coarse grooves and a high differential rate of speed. They are employed in the preparation of crêpe from bark-shavings and other sources where there is a large amount of impurity to be removed.

Drying.

The simplest method of drying the rubber is to hang up the strips of crêpe or sheets in a large and airy room. Light must be carefully excluded from the drying room, and thorough ventilation must be provided for. In the moist climates in which rubber is generally grown, drying is a very long process unless artificial means are employed for removing the moisture of the surrounding

Plate VIII

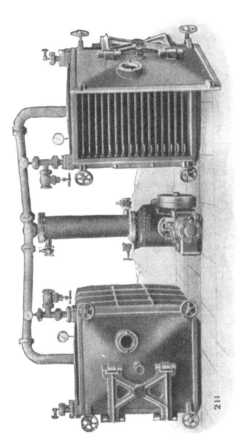

Vacuum Drying Machine

air. In practice the use of hot air is often adopted, the method being based on the processes customary in tea-withering lofts and in cacao-curing houses. Prolonged exposure to a high temperature is not however to be generally recommended, owing to the softening effect upon the rubber, and the danger that tackiness may arise. The use of cool air artificially dried has been sug-gested, such drying being effected either by mechanical cooling or by the use of such a substance as calcium chloride. However, the technical difficulties in the way of adopting such a process have not been entirely overcome, and the method is scarcely used in practice.

The most rapid and convenient way of removing the moisture is by the use of vacuum driers. In these machines the thinly crêped rubber is spread on trays in a square chamber or oven provided with a door which can be shut quite air tight. The chamber is heated by steam to a temperature of about 90 degrees F. It is then exhausted by means of a powerful air pump, and the pressure is kept low in this way for one or two hours. A large machine is capable of dealing with 200 to 300 lbs. of rubber in two hours. The rubber is taken from the drier in a soft and woolly condition, and is generally re-crêped whilst still warm. It should be pointed out that certain manufacturers are of opinion that rubber dried in this way is inferior in nerve to samples slowly dried in the open. Other experts how-ever consider vacuum-dried rubber to be as good as any other.

In fact, from the fluctuations in market price of the different forms of plantation rubber, it may be concluded that manufacturers have not yet arrived at a unanimous preference for any particular method of preparation. At successive auctions the market may be headed by pale crêpe, block rubber, smoked sheet or by some other form. From the planter's point of view, crêping and vacuum drying are probably the most convenient methods. It may therefore be useful briefly to recapitulate the successive processes gone through in an up-to-date factory in which these processes are employed. Although objections have been made to both these processes on account of their effect on the mysterious quality known as nerve, yet both are closely similar to some forms of treatment which the manufacturer himself employs. It seems probable therefore that objections to these methods will fade away as soon as large bulks which have been " worked " to an equal and known extent can be put upon the market.

Crêping and Vacuum Drying.

In a typical factory then, the spongy mass of rubber derived from the acidified latex is fed into the first washing machine. The rubber is passed repeatedly through this machine under a stream of water, and finally emerges as fairly thick crêpe. This crêpe is passed through a second machine in which the rollers are closely set, and is converted into a continuous strip of very thin or lace crêpe. The thin wet crêpe is arranged in thin layers on the trays of the vacuum drier, and when all are full the door of the oven is secured and the pumping machinery set to work. From the vacuum drier the rubber emerges after a couple of hours in a fluffy condition, looking very much like a blanket—much more so than the blanket crêpe into which it is next converted by passing repeatedly through a machine in which the rollers are set fairly wide apart. The crêpe is finally cut up into lengths convenient for packing in boxes, which contain from one to two cwts. of rubber.

An objection to crêpe, as usually exported at present, lies in the fact that it does not bear the brand of the estate impressed upon the rubber ; whereas sheet and block rubber is regularly marked in this way. A machine could easily be devised for stamping the sheets of crêpe at frequent intervals, and the use of some such method is strongly to be recommended.

Smoking.

It is not uncommon to combine the slow drying of rubber with a process of smoke curing, and for some time past rubber prepared in this way has commanded a higher price than the unsmoked variety. Many buyers believe that the creosote and other substances contained in the smoke exercise a preservative and strengthening effect upon the rubber. The curing of the wet rubber after sheeting or crêping is a process essentially similar to the curing of hams or of herrings. The strips of rubber are hung up in a drying chamber which is impregnated with smoke from a fire fed with green wood or coconut husks. It is usual to have the smoking house separate from the main factory in order to avoid danger from fire.

Smoking of Latex.

Various methods of preparing rubber by the direct action of smoke upon the latex have been suggested in imitation of the preparation of Hard Para rubber in Brazil; and a variety of machines have been devised for this purpose. Such methods of preparation are at present only tentative. Before they can be definitely recommended, further reports are required both on the quality of the rubber prepared in this way and on the facilities for adopting the method on a large scale in factories. It may be added that the last named necessity is one which inventors frequently seem to lose sight of.

In a machine devised by Mr H. A. Wickham the latex is allowed to flow upon the inner surface of a rotating cylinder, where it forms a thin film and is exposed to a jet of smoke obtained by burning coconut shells in a special stove. The rotation of the cylinder is so arranged that as each film of latex sets, another film is spread over its inner surface and is exposed in its turn to the smoke fumes. When a certain thickness of semi-solid rubber has been obtained, this is scraped out of the cylinder and pressed into a block. Rubber prepared in this way contains all the constituents of the latex, including a considerable percentage of water, and may be expected to resemble the Brazilian product closely. The chief point which still requires to be settled is how such rubber will compare in its physical properties with hard-cured Para on the one hand and with the usual forms of plantation rubber on the other hand. Rubber prepared by a somewhat similar method at the Singapore Botanic Gardens has been reported on very favourably by the manufacturers, but one or two experiments are not sufficient to establish the value of the method as compared with other processes already in operation on plantations.

Blocking.

Blocks of rubber are prepared in special presses by combining several layers of sheet or crêpe rubber, either smoked or unsmoked. Block rubber has the advantage of being very convenient for transport, and is also less liable to undergo damage during transit. From the point of view of the manufacturer, large blocks of rubber possess two disadvantages. It is necessary to cut up the blocks into smaller pieces before the rubber can be further dealt with in the factory. The presence of impurities or adulterations also cannot be detected simply by external inspection. Both these objections can be overcome by making flat blocks, which should not be more than one or two inches in thickness. Such blocks however are somewhat more troublesome to make than thick ones.

Scrap Rubber.

The highest grade of rubber prepared from the strained latex is known as First Latex Rubber. Other grades are prepared from the strainings of the latex, from the scraps of rubber which have dried on the bark of the trees or in the collecting cups, from the shavings of bark removed in paring, and even from the latex which has dried upon the ground beneath the trees. The scrap from small estates is sometimes sold as such, but on large estates the whole of it is turned into crêpe of various grades, according to the colour and the amount of impurity remaining after the washing process. After passing repeatedly through the macerating machines, the greater part of the impurities present are washed away, and the value of the rubber thus produced is only 20 or 30 per cent. less than that of First Latex Crêpe.

Packing.

The rubber is packed in wooden cases containing generally between 1 and 1½ cwts. It is most important that the inside of the cases should be smoothly planed and perfectly clean, and that no packing material of any kind should be used. Anything sticking to the rubber at once detracts from its value. The presence of the smallest amount of impurity may necessitate the addition of an elaborate process of purification prior to the other processes of manufacture. Just as in the case of tea and other products which are paid for by the pound, it should be the object of the packer, when making up the cases, to include such an amount of rubber as will weigh a few ounces more than an even number of pounds on arrival at its destination. This is owing to the fact that fractions of a pound are neglected in favour of the buyers, with the result that $100\frac{1}{4}$ lbs. are paid for as 100, whilst $99\frac{3}{4}$ lbs. are paid for as only 99 lbs. With rubber at four shillings a pound the matter is worth some attention. Some experience is required for determining just the right amount that should be included, owing to the fact that rubber is liable to lose a certain amount of weight in transport.

Sales and Markets.

The largest importing market of the world for rubber is New York, which receives nearly half of the total supply. Liverpool and London probably receive a comparatively higher proportion of the produce from Eastern plantations, although the value of plantation rubber is also beginning to be more fully recognised in New York. The other chief ports concerned in the rubber trade are Hamburg, Antwerp and Havre. In

New York sales are by personal contract for immediate delivery. In Liverpool and London, auction sales are held at stated intervals after samples have been exposed for a certain length of time. Private sales also take place in London, and a good deal of plantation rubber is now sold in advance by contract. Fortnightly auction sales of rubber are also held in Colombo and in Singapore.

The Best Form of Plantation Rubber.

We have seen that plantation rubber is placed upon the market in a variety of different forms, among which biscuit, sheet, crêpe and block are the most familiar. Any of these forms may be either smoked or unsmoked, and the unsmoked varieties may differ much in colour; whilst the thickness and other characteristics of the different types also vary considerably. It is a curious fact that whilst one of the principal demands of the manufacturer is for uniformity in the product which he buys, the present diversity is partly due to the failure of the manufacturers to make up their minds which type they prefer. In this way a vicious circle is established, and the desire of the producer to turn out a uniform product is partly frustrated by the buyer. On the other hand, it must be admitted that similar grades of rubber, when produced on different estates and from trees of different ages, appear to differ considerably in strength and resiliency.

Quality.

Light coloured rubber is valuable for certain purposes; for example transparent tubes for feeding bottles are now made from pale plantation rubber. The number of uses for which a very pale tint is necessary is however comparatively small. One obvious advantage of light coloured transparent rubber lies in the fact that its purity is unmistakable, as compared with dark commercial rubbers, and until recently a higher price was commanded by the palest crêpe rubbers. The quality of raw rubber, however, depends mainly upon what is known as nerve. This expression more or less sums up the results of tests upon the breaking strain, extensibility and resiliency, all of which should, generally speaking, be as high as possible. The resiliency of the rubber is measured by the pull exerted after a certain period of extension, and by the permanent extension shown after a certain period of stretching, followed by a certain interval of rest. The most resilient rubber exerts a large pull, and shows a small permanent extension. In practice, the strength and resiliency of the rubber are estimated by the brokers simply by handling and stretching; and although expert buyers arrive at a high degree of skill and judgment, such methods cannot be regarded as infallible. It is even whispered on estates that thick crêpe is preferred to thin on account of its greater apparent strength. In this particular form of rubber, strength is a feature particularly difficult to determine by rule of thumb methods. A simple and uniform method of testing commercial samples is therefore greatly to be desired. At the present time there is little doubt that differences in the original samples, and differences which arise during transport and storage, are often to some extent discounted by slight inaccuracies in the methods of testing. The determination of the best form of plantation rubber is thus further delayed.

It is therefore not surprising that the opinions of experts seem to differ as regards the precise effect of almost every factor which is capable of influencing the quality of the manufactured rubber. Thus rubber from young trees was formerly regarded as markedly inferior to that obtained from old trees. Recent experiments have not invariably confirmed this conclusion, although some appear to do so, and the question is not finally settled. In a similar way opinions differ widely as regards the effect of smoking or the use of vacuum driers. At present the general opinion is that plantation rubber is on the whole less satisfactory and less durable than fine Para. Thus it is commonly stated that plantation rubbers are unsuitable for the manufacture of elastic thread, a use which is generally admitted to represent one of the most stringent tests of quality.

As the crops on Eastern estates increase, and larger and larger bulks of rubber are turned out, a steady increase in uniformity is to be expected. Already many samples of plantation rubber have been produced which are indistinguishable in quality from the best Para by all ordinary tests. In fact, there is every reason for believing that in the future the highest plantation grades will share with hard Para the distinction of representing the highest standard of quality, even if they do not oust the American product from its present position of superiority.

As regards chemical purity, plantation rubber already surpasses any wild kind. It is therefore preferred by manufacturers to any wild rubber for the manufacture of rubber solutions and for waterproofing.

Defects and Blemishes.

Complaints are not uncommonly made of the appearance of spots and discoloured patches on the rubber turned out of the factory. In fairness be it remarked that such complaints are much less common than formerly. Petch states that the chief organisms concerned in the spotting of rubber biscuits appear to be bacteria and yeasts. Blood red spots sometimes appear. These have been attributed to the action of *Bacillus prodigiosus*, the microbe responsible for the so-called bleeding miracles. Brown and black patches may be due to the attacks of other bacteria. Discolouration may usually be prevented by attention to perfect cleanliness at every stage of preparation. All utensils which are brought in contact with the latex should be frequently scalded with boiling water, and the factory itself should be kept scrupulously clean. The hackneyed comparison with the conditions of an up-to-date dairy may profitably be borne in mind. The practice of frequent cleaning should extend to the tapping tools and collecting cups. It has been recommended that the latter should be of glass or earthenware, since proper cleansing is difficult with any form of metal cup. Glass or earthen cups are unfortunately very liable to be stolen on small estates.

The growth of mould fungi on the rubber sometimes causes trouble in wet weather. As a rule little damage is done in this way, the action being entirely superficial. Moulds seldom appear in large factories where the preparation is rapid and the storing and packing rooms are kept dry. Smoked rubber is practically immune from the attacks of moulds, owing to the antiseptic properties of the smoke constituents.

Tackiness.

Rubber is said to become tacky when the surface turns soft and sticky. In some cases the softening may proceed so far that whole sections of rubber fall to pieces and dissolve into a liquid form. Tacky rubber is useless for the ordinary purposes of manufacture, and can only be sold at a low price as a subsidiary product. The condition is rightly dreaded as the worst defect which can possibly arise.

One certain cause of tackiness is exposure to sunlight, and it is to guard against tackiness that drying and storing rooms must be carefully guarded from excessive light. According to Petch, the assertion that tackiness can also be caused by the action of bacteria has not been conclusively proved. It is quite possible however that tackiness exists in different forms, and may be due to different causes. The use of too strong acid in coagulation may lead to tackiness, and it is generally believed that the condition arises more readily if the acid employed in coagulation is not thoroughly washed out of the rubber. Heat has also been mentioned as a cause, although the softening due to a high temperature is usually temporary, and passes off when the rubber is cooled. The fact is that the conditions which lead to tackiness are by no means fully understood, and further study is desirable.

CHAPTER VIII

THE PESTS AND DISEASES OF *HEVEA*

Plantation Conditions.

PROBABLY no species of plant is exempt from the attacks of some kind of animal or vegetable enemy or parasite. Under natural conditions, however, it is rare to find any disease developing the proportions of an epidemic and killing off large numbers of plants in a particular area. In the case of tropical trees like *Hevea*, one very good reason for the absence of epidemics under the ordinary conditions of forest growth is readily discernible. Unlike the uniform woods characteristic of temperate climates, a tropical forest almost invariably consists of a varied mixture of different species of trees. This mixed character of tropical vegetation extends so far that it has been estimated that as many as three hundred different species of trees are often to be found on a single acre of ground, whilst two individuals of the same species practically never stand side by side.

The conditions obtaining in a plantation are precisely the reverse of those natural to a tropical forest.

Hundreds of thousands of trees of the same species are here arranged in regular rows, their leaves and roots actually intermingling. The conditions are therefore ideal for the development into an epidemic of any disease which may make its appearance.

Epidemics.

Consequently the occurrence of epidemics under the conditions described is by no means unknown. One of the most famous examples of modern times is afforded by the coffee leaf fungus *Hemileia vastatrix*, which, in combination with other diseases, entirely ruined the coffee planting industry of Ceylon in the seventies and early eighties. This calamity has taught planters in general a lesson which it is to be hoped they will not readily forget; for if a disease be taken in hand in its early stages, when only a few individuals are affected, it is generally possible to cope with the trouble to some extent and to prevent its universal spread. With this object in view, superintendents are now instructed to keep a close look-out for the first indications of disease; and expert plant-doctors—entomologists and mycologists—are maintained in all important tropical planting countries. It is their business to diagnose correctly any symptoms of disease which may appear, and to prescribe the most appropriate remedies which science has been able to devise.

And, just as in the case of human ailments, the

general practitioner is relied upon for the treatment of well-known complaints, but the scientific specialist is called in where the cases are difficult to diagnose, so it should be in the case of plant diseases. For everyday work the planter is his own doctor, and a good planter should be familiar with the nature and treatment of the common diseases to which his crop is liable. The recognition of new diseases and the devising of appropriate remedies calls for scientific research by highly qualified specialists. No community of planters can regard itself as safe from the danger of new diseases which does not maintain a properly equipped establishment for research, headed by a scientific officer of the highest possible qualifications. Probably in no branch of commercial enterprise does the liberal treatment of science pay better than in planting.

Hitherto no disease of *Hevea* has assumed the proportions of a serious epidemic, and we have no reason for anticipating that a fate similar to that of coffee in Ceylon awaits this industry in any country where rubber is now cultivated. Minor ailments exist however in some variety, and it behoves the planter as well as the agricultural official to familiarise himself with the known symptoms of disease and with the simpler recognised remedies. He can then take immediate steps to prevent the wide extension of trouble from small beginnings; serious trouble is only likely to occur if unhealthy individuals are neglected. As prevention is better than cure, an elementary knowledge of the general principles of plant sanitation is even more important than a knowledge of special diseases.

Wind.

Perhaps the most serious of all pests which affect the health of rubber plantations is an inanimate one, namely wind. In regions which are severely wind-swept it is useless to attempt the growth of *Hevea* rubber. In less exposed situations the force of the wind may be broken by leaving belts of jungle when the original clearing is carried out, or by planting rows of hardy and quick-growing trees as special wind-breaks. The idea of jungle belts has frequently been suggested as a means of checking the spread of fungus diseases, but it appears to have been very seldom carried into effect, the planter preferring to occupy the whole space at his disposal with his own proper crop. Such belts moreover possess this serious disadvantage; that if they are wide enough to prevent the passage of wind-borne fungus spores they are also large enough to afford an excellent harbourage for weeds and for many of the vertebrate enemies mentioned in the next paragraph.

Animals.

Among the larger animal pests of rubber plantations are elephants, deer, cattle, pigs, monkeys and porcupines. All these animals are liable to effect widespread destruction in young clearings, and porcupines may even cause serious damage to adult trees by gnawing off the bark. A good deal can be done by fencing to exclude all the above-named animals with the exception of monkeys, but in order to be effective the fencing must be very thoroughly carried out. Deer can easily clear a six-foot fence, whilst barbed wire is a matter of supreme indifference to an elephant who has made up his mind to proceed in a certain direction. The lower part of the fence must be of wire netting if porcupines are to be excluded.

Insects.

Under ordinary conditions a healthy *Hevea* tree is practically immune from the attacks of boring insects, owing to the presence of latex, which is found an effectual check to the progress of these pests. If, on the other hand, an area of bark is killed by wounding or by the attacks of fungi, the latex soon dries up. Boring beetles and other insects then readily obtain access, and contribute materially to the damage done to the tree. The insect attacks are however purely secondary, and their prevention is essentially bound up with that of the original cause of damage, which is usually a fungus disease.

There is however one insect, prevalent in the Malay Peninsula, which is able to cause considerable damage to living *Hevea* trees without the assistance of fungi. This is a species of white ant known as *Termes Gestroi*, which is apparently able to effect an entry through the roots of entirely healthy trees. The insects then rapidly proceed to eat out the interior of the wood. *Termes Gestroi* is a pest which requires to be seriously taken in hand at its very first appearance. By digging a shallow trench round an affected tree it is often possible to locate the position of the termites' nest, because the insects quickly build covered ways across the bottom of the trench in order to obtain access to their food supply, and by following these up the direction of the nest can be determined. The best means of exterminating the white ants is by pumping into their nests the fumes of sulphur and white arsenic, generated by heating these substances over burning charcoal in a special apparatus. The channels excavated in the affected tree may be similarly fumigated, but if the damage has proceeded far it is simpler and more effective to cut down the whole tree and burn it.

In its early stages of growth *Hevea* is susceptible to the attacks of a somewhat larger variety of insect pests. Young plants in nurseries, and those which have recently been planted out in clearings, are subject to damage from cockchafer grubs, cutworms and similar subterranean creatures which gnaw the roots. These may be got rid of to some extent, at least in nurseries, by treating the soil with kainit or with " vapourite," a preparation sold for the purpose.

One animal against which the presence of latex is no protection is a slug—*Mariaella Dussumieri*—which actually drinks the latex, and appears to thrive upon a beverage which most animals would find decidedly indigestible. These slugs may do considerable damage by eating off the buds and young leaves. They may be prevented from climbing the trees by painting a band of tar round the trunk. Another method is to steep sawdust in a ten per cent. solution of carbolic acid, and sprinkle it round the bases of the trees. Such measures, in addition to protecting the trees for the time being, soon lead to a diminution in the numbers of the pest by cutting off the latter from its principal supply of food.

Fungus Diseases.

On the whole, the enemies of *Hevea* belonging to the animal kingdom are less deadly than the vegetable pests reckoned among the parasitic fungi. The most important, that is to say the commonest, diseases of *Hevea* which are due to the attacks of fungi, are five in number. These may be divided into two groups, according to their point of attack. Two of the common diseases attack the roots, whilst three of them are found upon the trunk or branches. One of the most serious of the latter also attacks the fruits. In addition to these, there are a number of other diseases which are of smaller importance from the planters' point of view, either because of their rarity or because the damage hitherto ascribed to their attacks is insignificant. Such diseases are however of special interest to the professed mycologist, whose business it is to study the conditions under which they might become more prominent, with a view to ensuring, if possible, that the necessary conditions do not arise. Some of the fungi in question attack the roots, others the stems and others again the leaves, especially in the case of young plants.

Diseases of the Roots.

Root diseases are particularly dangerous, because their early stages generally pass unnoticed, and very often the first indication of their presence is given by the death of the tree. Sometimes the main tap-root may be entirely destroyed without any sign of injury appearing above ground, since sufficient supplies of water and salts are provided by the lateral roots to maintain the crown of foliage in a condition of apparent health. One day a storm of wind brings down the whole tree, and the full extent of the injury is disclosed. Occasionally a single lateral root may be discovered showing symptoms of disease. In such a case it may be possible to cut away the diseased portion and to save the remainder of the tree. In the majority of cases, however, the planter must make up his mind to destroy the affected tree in the hope of saving its neighbours.

Altogether, three different root diseases are at present recognised, which may be readily distinguished from one another on examining the affected roots. In the case of the fungus known as *Fomes semitostus*, which is the most serious of the three, the dead root is found to be covered with white or yellowish threads and cords of *mycelium*—the growing strands of the fungus. An equally common, but not quite so destructive, fungus is *Hymenochaete noxia*, which may be recognised from the fact that the root killed by it is covered with an encrustation of sand and small stones, cemented together by the brown or black mycelium. There is a third less common fungus which is occasionally found upon the roots of *Hevea* in Ceylon, known as *Sphaerostilbe repens*. When this is present there is no external mycelium on the surface of the root, but on peeling off the bark, dark brown strands of mycelium are found wandering over the surface of the wood.

Fomes semitostus.

The fructification of *Fomes semitostus* does not usually make its appearance until long after the death of the tree. Where rubber trees have died however it is frequently to be found on a neighbouring jungle stump. The fructification or *sporophore* grows out in the form of a flat semi-circular bracket from the decaying log or stump. It is distinguishable from innumerable other " bracket fungi " by its characteristic colours. When fresh, the upper surface is coloured a rich brown with a narrow rim of yellow, whilst the under surface is bright orange. The consistency of the fructification is woody and brittle, and the lower surface is covered with enormous numbers of very minute holes. The latter represent the openings of the tubes in which the spores are produced.

According to Petch, the disease makes its appearance as a rule when the plantation is from one to three years old. This is due to the fact that the fungus spores do not attack the living trees directly. The spores, however, germinate readily on dead jungle stumps remaining in the soil, and develop a vigorous mycelium. Stout strands of mycelium, known as *rhizomorphs*, then grow out through the soil, and are able to attack the living roots of *Hevea*.

When the presence of this disease is recognised in a plantation—generally owing to the blowing down of one or more trees—immediate steps must be taken to destroy the affected trees and to prevent further loss. The roots of all affected trees must be dug up and burnt, and in particular the jungle stump in which the evil originated must be found, dug out and thoroughly destroyed by fire. When the limits of the group of affected trees have been ascertained—if necessary by digging down to examine the roots—a trench from 18 inches to two feet in depth should be dug right round the affected area. The earth removed in digging should be thrown inside the trench. The ground enclosed within the trench should be deeply dug all over and lime forked in, and the bottom of the trench, which must be kept clear of weeds, should be covered with the same material. Petch does not recommend replanting with *Hevea* in less than twelve months from the date of destroying the diseased trees. This method of isolating an infected area by trenching was first proposed by Hartig for use in forestry work. It was recommended by Massee for dealing with *Rosellinia* in tea.

Hymenochaete noxia.

The fructification of *Hymenochaete noxia*—commonly known as Brown Root Disease—is much less conspicuous than that of *Fomes*, and is also comparatively rare. When present it takes the form of a thin brown crust adhering to the base of the trunk. The surface of the crust is velvety, being covered with almost microscopical bristles. The disease is not confined to *Hevea*, but is also found on *Castilloa*, tea, cacao, camphor, coca and the shade tree *Erythrina lithosperma* (dadap) in Ceylon ; on *Funtumia* on the Gold Coast and on coffee in Java. The majority of cases affecting *Hevea* in Ceylon have occurred where old cacao has been cleared in order to plant rubber. Unlike *Fomes*, the fungus does not appear to travel independently through the soil, but is only transmitted where the roots are in contact with other diseased roots or with dead wood upon which the fungus is growing. In order to prevent the appearance of the disease, the stumps of cacao which have been cut down in order to make way for rubber should be carefully extracted. *Hevea* trees killed by the fungus should be cut down and the roots dug up and burnt. The soil which the roots of the dead tree occupied should be dug over and quicklime forked in, but the elaborate precautions recommended in the case of *Fomes* do not appear to be necessary, as the fungus seldom spreads after the source of infection has been removed.

On the other hand, if *Sphaerostilbe* makes its appearance, the treatment recommended for *Fomes* should be carefully carried out, since this fungus, too, is able to spread through the soil to adjacent trees by means of a free-growing mycelium.

Diseases of the Stem—Canker.

Three diseases of the stem are more or less prevalent in *Hevea*, whilst four others which at present appear to be of minor importance have been recorded from time to time.

Probably the most serious of the stem diseases is the so-called canker, which has recently been shown to be due to the attacks of the fungus *Phytophthora Faberi*, and is therefore a near relative of the deadly potato disease. Like the potato disease, the canker of rubber is disseminated by motile spores, which swim actively in the film of water which may cover the surface of the trunk in wet weather. The canker of rubber has been shown to be identical with that of cacao, and the same fungus attacks the fruits as well as the bark in both species. The term canker is a singularly unsuitable one for this disease, which gives rise to no roughness or eruptive outgrowth such as is usually associated with the name of canker. The disease consists in a softening and rotting of the inner layers of bark, which may be scarcely visible from the exterior, or may betray its presence by the oozing out of a dark red liquid. The softened and discoloured area is occupied by the hyphae of the destructive fungus. The rotting bark may soon become the breeding ground for other fungi, but the *Phytophthora* is the primary cause of the mischief. On shaving off the outer layers of bark with a sharp knife, the diseased area becomes recognisable as a brownish or claret coloured patch, with well defined edges separating it from the surrounding healthy bark.

The only remedy is excision. If the disease is discovered whilst the patch is still small, the whole of the diseased tissue can be cut away. The healthy bark should be trimmed all round the exposed area of wood with a very clean cut, and if much wood is exposed the latter should be tarred. In the course of time the bark will heal over the wounded area and the tree will recover. The disease is specially serious, owing to the fact that it generally attacks the lower part of the trunk where tapping is in progress. Tapping must therefore be suspended until the recovery of the tree is well advanced. In cases where the disease has already spread far round the circumference, it is better to cut down the tree at once. Any diseased portions cut away, and, in the case of felling, the whole tree, should always be destroyed by fire.

Plate IX

Photo T. Petch

Canker of *Hevea* Bark

The same fungus attacks the fruits of *Hevea*, especially in wet seasons. The pods turn black and sodden, and rot upon the trees. Beyond the loss of the seed crop, little harm appears to be done directly to the trees, but the diseased pods serve as centres of infection, and thus increase the chance of the disease developing upon the bark. The only possible treatment is to collect and burn the diseased fruits.

The spread of canker is greatly facilitated by the shade consequent upon close planting. In districts where the disease is prevalent it is therefore desirable that the trees should be widely spaced. Where they are already closely planted, the question of thinning out becomes a serious one, but if it is decided to fell a certain proportion of the trees, the stumps and larger roots should also be extracted in order to avoid danger from root disease. It is understood that experiments are being tried in Ceylon to find out whether the tapping area can be sprayed with some substance which will prevent the germination of the fungus spores, whilst not injuring the quality of the rubber prepared from the latex. Such spraying is not likely to be of much use so long as rain water is allowed to flow over the surface of the tree, whilst, if the trunk can be kept dry, the germination of spores will probably be checked as effectively as by a fungicide. The provision of rain guttering above the tapping area seems therefore to be a most desirable measure in all districts where rubber canker is prevalent.

Pink Disease.

The malady known as Pink disease, due to the attacks of the fungus *Corticium salmonicolor*, has much more the appearance of what is commonly regarded as a canker. This disease makes its appearance as a pink incrustation covering the bark. The patch may extend until it covers a large area, in the centre of which the bark is dead and dry, whilst at the edges the advancing fungus is only superficial. Spores are formed upon the surface of the pink patch, and are carried by the wind until they find a lodgement upon the surface of the bark. This commonly occurs at a point where the trunk forks into two or more branches, and it is in such a position that the disease generally arises. Hyphae from the germinating spores penetrate into the bark and destroy its living tissue. From the centre thus established the disease then spreads widely over the surface.

The disease is readily recognisable owing to the pink colouration which it produces. When the fungus makes its appearance upon the upper branches of a young tree, the diseased branches may be cut off completely and burnt. On older trees, as in the case of canker, it is sometimes possible to cut out the diseased patch of bark, if the presence of the fungus is recognised at a sufficiently early stage. In South India the application of *Bordeaux mixture* is widely adopted as a preventive measure. The fungicide is simply painted on to the trees with a brush. This remedy is found to be effective, and the cost of application is small in comparison with the value of the immunity obtained. Bordeaux mixture is made by dissolving 5 lbs. of copper sulphate in 25 gallons of water in a wooden barrel, and pouring this solution into a large vessel simultaneously with a like amount of milk of lime. The mixture must be continuously stirred during the process. The milk of lime is prepared by pouring water very gradually upon 5 lbs. of quicklime, with constant stirring, the slaked lime being finally made up to 25 gallons. Theoretically, the amount of lime should be rather less than that of the copper sulphate, but in practice perfectly pure quicklime is not obtainable, so that some margin may be allowed.

Die-Back.

A third disease which attacks the leading shoots of *Hevea* is known as die-back. This expression describes the nature of the disease with considerable accuracy. The fungus responsible for the disease is known as *Gloeosporium alborubrum*. This fungus is microscopic, and its presence can only be recognised from its results. These, however, are not generally serious unless the dead shoots are attacked by another fungus, *Botryodiplodia theobromae*, which, as a rule, cannot by itself attack living shoots.

When only the first named fungus is present, the dying back of the stem is confined to the leading shoots; fresh shoots grow out from buds lower down the stem to take the place of the dead branches, and but little damage is done. But when the *Botryodiplodia* obtains a footing, the disease pursues its downward way with great rapidity, and ultimately destroys the whole tree. Here, again, the only remedy is to cut off and burn the diseased portion. This should be done with a clean cut some distance below the lowest signs of injury, and the cut surface should be tarred.

Botryodiplodia has also been known to attack young stumps shortly after planting. In such cases the fungus probably effects an entrance through injuries in the stem or upper part of the root. The holes in which the trees are planted should in such cases be treated with lime before supplying with fresh plants.

Burrs and Nodules.

A common complaint affecting the stems of *Hevea* trees, to which no sufficient cause has yet been assigned, consists in the development of woody nodules in the bark. These sometimes make their appearance upon a large number of trees at particular periods in a manner highly suggestive of an epidemic, and grow so rapidly as to interfere seriously with the operation of tapping. Hitherto it has not been found possible to show that the nodules are associated with any specific organism attacking the bark; nor are they constantly associated with mechanical injury, although there is evidence to

Plate X

Photo H. F. Macmillan

Nodules in *Hevea* Bark

show that they appear in greater numbers on trees which have been tapped with a pricking instrument than on trees which have only been pared. In fact, the whole subject of their origin seems to require further investigation.

The nodules arise as minute woody bodies buried in the growing bark, quite independently of the proper wood of the tree. Their presence is first betrayed by a swelling and cracking of the outer bark. If the surface of such a swelling is sliced off with a sharp knife, numerous nodules, about the size of small peas, are found embedded in the bark. At this stage they may be carefully extracted with the point of a knife, and if this extraction is carried out with thoroughness the bark recovers and no further injury is done. If neglected, however, the nodules grow with great rapidity, and fuse with one another and with the actual wood of the tree. The surface of the tree now shows extensive rugged prominences over which tapping by the ordinary process of paring is no longer possible. The woody mass may still be prized off with a hatchet or a crow-bar, but a ragged and extensive wound is left which takes a long time to heal.

At certain seasons these nodules have appeared on large numbers of trees all over Ceylon. At other times their appearance may pass almost unrecorded for years together. It was observed that the worst visitation occurred after a prolonged period of drought, and was most severe at high elevations and on trees which had been severely tapped, especially where the pricker had been employed. The appearance of the nodules is therefore apparently facilitated by anything which tends to weaken the vitality of the trees.

The outbreak in question occurred in the spring of 1911. At this time a marked difference in the severity of the epidemic was observed on two similar plots of *Hevea* trees at Peradeniya, each one acre in extent. On plot A upwards of 60 per cent. of the trees were affected, and on plot B only about 10 per cent. Plot A had received during the two preceding years applications of a concentrated mixture containing 150 lbs. of sulphate of ammonia, 100 lbs. concentrated superphosphate and 100 lbs. sulphate of potash. Plot B was unmanured. During the twelve months immediately preceding the outbreak, plot A had been tapped by paring and pricking with the spur-shaped pricker, whilst the trees on plot B were pared only—on a similar system. It seems natural to associate the larger proportion of nodular outgrowths on plot A rather with the pricking than with the manuring, but it is apparent that the whole subject requires further study.

General Sanitation.

In a well lighted plantation which is kept clear of decaying stumps and branches, fungus diseases are not likely to make their appearance except sporadically. Close planting and the presence of intercrops produce conditions favourable for the introduction and spread of stem and leaf diseases, whilst decaying stumps remaining in the soil constitute a fruitful source of fungus diseases of the roots. Petch, in fact, goes so far as to state that if there were no dead stumps there would be no root diseases either in *Hevea* or tea. The conclusions to be drawn from these facts are obvious, but they apply mainly to the precautions which ought to be taken during the first opening up of the estate. We have now to consider the sanitation of an estate in bearing.

On well managed plantations a special gang of labourers is often employed, whose business it is to make the circuit of the estate and to keep a close look out for the appearance of disease. This periodical inspection is specially necessary in districts subject to the attacks of die-back and pink disease, since these maladies may readily escape the attention of the tapping coolies. In such districts the removal of dead branches should constantly be carried out as fast as death overtakes them, and by this means the spread of the diseases may be reduced to a minimum. The pruning of lateral branches should always be done by means of a clean cut flush with the surface of the parent stem, in order that no projecting portion may be left to die back and offer a point of entry for disease germs. All extensive areas of exposed wood should be covered with tar. For this purpose ordinary gas tar is preferable to Stockholm tar. All branches and debris removed in pruning should at once be burned.

For dealing successfully with plant diseases, intelligent cooperation is eminently desirable. A single neglected and diseased property may constitute a serious menace to the health of the products on all the other estates in the same district. The case of an owner who is not amenable to the principles of cooperation in the treatment of diseases would constitute a strong argument in favour of coercive legislation. For the owner who neglects to take ordinary precautions for preventing the spread of disease is interfering with the rights of his neighbour almost as much as if he were to set fire to a corner of his plantation.

CHAPTER IX

THE CULTIVATION OF SPECIES OTHER THAN
HEVEA BRASILIENSIS

ALTHOUGH *Hevea brasiliensis* is universally recognised as the plantation rubber tree *par excellence*,
several other kinds of rubber-producing plants are used
extensively for the same purpose. In spite of its
hardiness, the adaptability of *Hevea* is limited, and in
dry climates *Manihot Glaziovii* is probably to be preferred. In fact, extensive plantations of the last named
species have already been opened in the drier regions
of Africa as well as in its native country, the Ceara
province of Brazil. In the moist climate of West Africa
Funtumia is said to be quite as successful in plantations
as *Hevea*, although the evidence upon the subject is
somewhat restricted. Finally, in Mexico large plantations of the indigenous *Castilloa* have been established.
Here, however, it is doubtful whether the planters would
not have been better advised to introduce *Hevea*, since
the climatic conditions suitable for the two trees are
closely similar.

Castilloa.

As stated in Chapter I, plants of *Castilloa* were brought successfully to Kew prior to the arrival of the most important consignment of *Hevea*, and trials have been made with the former plant in many parts of the tropics, including almost every British tropical colony. So far as we are aware, *Castilloa* has nowhere been found equal to *Hevea* in suitability for plantation use, in spite of fairly exhaustive trials in many different districts.

The home of the *Castilloa* tree is Mexico and Central America. The total export of rubber from Mexico in 1908—1909 was about 1,000,000 lbs. Of this amount, some 40 per cent. is stated to have been plantation rubber. In 1910 the area of rubber plantations in Mexico is believed to have been about 90,000 acres. The greater part of this area was planted between 1897 and 1906. The capital involved, mostly subscribed in the United States, was estimated at about £4,000,000.

So far, these plantations have not by any means fulfilled the expectations of their promoters. A yield of 67 lbs. of dry rubber per acre from six- to eight-year-old trees, although gathered in at a trifling cost, is very small compared with the yield from the plantations of *Hevea* in the East. The latex from young trees, moreover, contains a high proportion of resin, and the period of waiting for satisfactory crops is consequently a long one.

The *Castilloa* has generally been closely planted. From 200 to 300 trees per acre is by no means an uncommon estimate. Since the growth of *Castilloa* in the rich soils of Mexico appears to be considerably more rapid than the average recorded for *Hevea* on plantations, there can be little doubt that the trees are greatly crowded.

The trees are only tapped at intervals of about four months. The method employed is purely one of incision, and consists in cutting a few slanting or V-shaped channels in the bark. The rubber is usually prepared from the latex by a process of creaming. In this process the latex is diluted with a considerable bulk of water and, after stirring, is allowed to stand in tanks until the whole of the rubber has collected as a cream upon the surface. The clear liquid is then drawn off from below, and the whole process is repeated until dirt and soluble impurities have largely been removed. The rubber is then finally dried and rolled into sheets. Centrifugal machines are also sometimes employed. In this way much time can be saved, and the washing is more completely carried out.

In Ceylon very fine samples of smoked sheet rubber have been prepared from *Castilloa* latex. The cultivation, never very extensive, has however been almost entirely given up in favour of *Hevea*. In the West Indies *Castilloa* is still under trial, but only on a small scale, and the amount of land available for the cultivation is limited. In Jamaica a yield of 200 lbs. per acre is anticipated from *Castilloa* after the tenth year. This form of rubber has also been planted somewhat widely in German colonies, particularly in New Guinea.

In spite of the small yields per acre, *Castilloa* is still regarded by some writers as being the most suitable form of rubber for plantation purposes in Mexico, owing to the small cost of collection. Although the total yields are small, the yield from a single tapping is very much larger than in the case of *Hevea* at the same age. The cost of labour is very much higher in Mexico than in either Africa or Asia.

Manihot.

Ceara rubber, *Manihot Glaziovii*, was the third species to be widely distributed to British possessions in the Tropics during the seventies. Introduced to Kew from North East Brazil by Cross in 1876, seeds or plants were sent out to most of the tropical colonies during the following year. In Ceylon this species attained a certain amount of popularity some years before *Hevea* came to be at all widely planted. *Manihot* was planted in the Trincomalee district in 1880, where, however, in spite of the richness of the soil, planting has made but little headway. Nevertheless, in 1883, no less than 977 acres were under cultivation with this product in Ceylon.

Difficulties of tapping, which have not even now

been entirely overcome, soon led to a temporary cessa-
tion of planting, and after 1884 the interest in *Manihot*
died away, owing to the small yields of rubber obtained.
Samples of rubber from Zanzibar were unfavourably
reported on in 1884, and further extension of the
cultivation of *Manihot* in Africa was thus delayed.
Recently, however, the species has been widely planted
in East and Central Africa and in Angola, and good
results are anticipated, owing to the fact that Ceara
rubber grows well in a drier climate and at a higher
elevation than *Hevea*.

Biffen, who visited the North East coast of Brazil
in 1897, found large plantations already being opened
in the Ceara district, at elevations up to 3500 feet.
No recent reports regarding the development of these
plantations are available. The methods of collecting the
latex and preparing the rubber on plantations in Brazil
appear to be similar to those applied to the wild trees.
Tapping is generally carried out during the dry season
by slicing the bark with a knife.

The Ceara rubber plant grows readily from either
seeds or cuttings, but in order to ensure immediate
germination of the seeds it is necessary either to file
through the tip of the hard shell or to steep the seeds
for some time in warm water. The early growth of the
trees is very rapid, and it is often possible to begin
tapping earlier than in the case of *Hevea*. Close
planting is generally recommended, but this seems to be
a mistake. Trees have been planted at distances of

15 by 15 feet in more than one district in Ceylon, and
growth has been very rapid for the first two years.
At the end of this time the branches met and formed
a close cover, with the result that subsequent growth
was very slow.

The structure of the laticiferous system of *Manihot*
is closely similar to that of *Hevea*. The *Manihots* are
the only other rubber-producing plants upon which
tapping can be carried out at an equally short interval,
and although the study of the subject has not been
carried so far as in the case of *Hevea*, the phenomenon
of wound-response seems to be closely similar in both.

Another advantage of Ceara rubber is that the plant
grows like a weed, so that trees which have been injured
in the course of tapping can be replaced with very little
trouble.

On plantations, the hard outer bark is generally
removed before tapping. Opinions differ as to whether
tapping should take place immediately after the outer
bark is removed or whether it should be delayed for
a longer or shorter period. In Africa the method of
tapping usually adopted is the primitive one of pricking
or stabbing with a blunt thin-bladed knife. The pricks
are made about an inch apart in a vertical series. Sometimes the bark is first rubbed over with a freshly cut
lime or lemon, in order that the latex may be coagulated
on the bark by the citric acid thus applied. The strips
of wet rubber formed upon the bark are collected by
rolling them up on sticks or small wooden rollers.

When a thickness of about a quarter of an inch of rubber has been obtained, the rolls are slit up so as to form small sheets, which are subsequently washed and dried.

The method of alternate paring and pricking recommended in Hawaii has been found to cause considerable damage to the trees in Ceylon. Paring on the herringbone system has sometimes been found to give good results if not carried too far. If the paring extends over only an inch of bark, and a space of an inch is then left untapped before the next cut is made, the bark is generally found to heal satisfactorily. Rather wide shallow paring with a small gouge has given good results on Ceylon estates, and has been followed by perfect renewal. In many climates tapping can only take place during dry intervals in the wet season, since the trees drop their leaves during the dry season of the year, and then yield little or no latex.

Coagulation takes place on the simple addition of water. Very good samples of rubber, closely comparable with the best *Hevea* sheet in appearance, have been prepared by diluting the latex with an equal bulk of water and allowing it to stand until coagulation is complete. The further processes involved are precisely similar to those adopted in the preparation of *Hevea* sheets or biscuits.

A process which should be rigidly avoided is the mixing of the latices of *Manihot* and *Hevea* on estates where both products are cultivated. Such mixture leads to rapid coagulation without reagents, and to the production of a rubber which is excellent in appearance. Rubber of this kind is not, however, desired by manufacturers, and the mixed character of the sample is easily detected by experts. *Manihot* rubber requires slightly different treatment in manufacture from the produce of *Hevea*, and the mixed product is therefore unsatisfactory.

The yields from young trees appear to be similar to those from *Hevea*; in subsequent years, however, the increase in yield is not so great. In Nyassaland in 1910 four hundred four-year-old trees are said to have yielded an ounce of dry rubber apiece from only two tappings, and the cost of collection was only 4*d*. per lb. In this case the tapping consisted of making vertical rows of pricks. At Peradeniya, Ceylon, experiments in paring were carried out on trees only three and a half years old at the beginning of the experiment, planted at the rate of 200 to the acre. About four ounces of rubber per tree were obtained in a year by 70 tappings. A yield of nearly 50 lbs. per acre was thus obtained at an earlier age than that at which a similar amount of rubber could safely be extracted from *Hevea* trees. The cost of production, however, was high.

Among other species of *Manihot*, *M. dichotoma* was introduced into Ceylon in 1908. So far, none of the new species have succeeded so well as *Manihot Glaziovii*. The growth of the trees and the thickness of bark were poorer at equal ages, and in the few experiments made the yield of latex was less. *Manihot dichotoma* also suffered severely from the effects of wind of no great force ; the trees were uprooted and branches were split off in every direction.

Funtumia.

Funtumia elastica and its cultivation in Africa have been made the exclusive subjects of a recent book by C. Christy (*The African Rubber Industry and Funtumia elastica*). From this work the few remarks here given are partly summarised, and for fuller information reference may be made to the original source.

Some description of the species itself was given in Chapter II of the present work. The seeds retain their vitality well. They are very small, with a plume of long silky hairs, and one ounce of seed is sufficient for 20 square yards of seed bed. As germination at even distances cannot be guaranteed, the seedlings may with advantage be transplanted to a second seed bed prior to planting out in the field. The further operations of planting and cultivation are similar to those adopted in the case of *Hevea*.

Very close planting is recommended, to be followed by subsequent thinning. Close planting is said to be necessary in order to obtain straight clean stems, to obliterate lower branches which would interfere with tapping, and in order to minimise the cost of weeding. The thinning subsequently undertaken gradually leads to a distance of 12 by 12 or 15 by 15 feet at six or seven years of age. The trees are deep rooting; consequently manuring and cultivation have less effect upon growth and yields than in the case of *Hevea*. A fair yield is obtained from the sixth year onward. The average growth in plantations is said to be of the same order as that of *Hevea*, the increase in girth being from three to four inches per annum.

As in the case of *Castilloa*, the trees can only be tapped two or three times in a year, and great care has to be exercised in order to avoid damage to the bark. The method of tapping recommended by Christy is by cutting shallow conducting grooves on a half-spiral system. The slanting grooves are then to be pricked with a thin-bladed spur pricker. Eight inches is regarded as the best distance between the lateral channels.

The trees are first tapped during the sixth year. They are then tapped twice a year until they are eight years old, and afterwards three times a year. The later cuts are made about two inches above the old, and a fresh vertical channel is made at each tapping. On old trees the tapping may be carried to a height of 30 feet.

In a particular series of experiments, about five ounces of dry rubber per tree was obtained by tapping to a height of 30 feet on the double half-spiral system. The average girth of the trees concerned was 28 inches. Christy gives the following estimates of the average annual yield per tree under plantation conditions.

Age, years	...	...	5	6	7	8	9	10
Dry rubber, ounces	...	3	4	5	9	12	15

Coagulation is effected by simple dilution with water or by the use of various reagents, amongst which tannic acid, formalin and hydrofluoric acid are said to be specially effective. The fresh latex is neutral in reaction or very slightly acid, and acetic acid alone does not produce coagulation. Centrifugal machines are also useless, probably on account of the very small size of the latex globules. The physical properties of the rubber are found to vary considerably, according to the nature of the chemical reagent employed.

Christy regards *Funtumia* as better suited than *Hevea* for growth in African plantations. It is less liable to the attacks of insect pests, and is much cheaper to cultivate. It is also considerably cheaper to tap, owing to the relatively large yields obtained at a single tapping. Plantations have already been opened on an extensive scale in Kamarun, and the cultivation is under trial in Southern Nigeria and in other parts of British West Africa.

The introduction of *Funtumia* into countries outside Africa has not generally been attended with much success, for in other climates the plant appears to suffer widely from the attacks of insects. At Pera-deniya the growth of the trees was found to be very slow, largely owing to the depredations of a leaf-rolling caterpillar, which led annually to the complete defoliation of every branch.

Ficus elastica.

Sir Daniel Morris, in his Cantor Lectures, states that
" In 1873 the Government of Bengal decided to start
regular plantations of *Ficus elastica* in Assam. The
order, issued in 1873, was repeated in 1876, and has
been acted upon with slight interruption until the
present time." In 1884 nearly 900 acres had been
planted in the Charduar district, and in the same year
directions were issued that the plantations should be
increased by 200 acres annually. It is said to be well
known that although the trees grow vigorously in
situations remote from the hills, the yields of rubber
are then almost negligible. It appears, however, that
even in its native hills the yields of the Assam india-
rubber tree are very small from a planter's point of view.
Thus in 1896 the Inspector General of Forests, H. C.
Hill, was glad to estimate a yield of a maund (about
80 lbs.) per acre at an age of 50 years.

Ficus elastica has been widely planted in Java in
more recent years, but here again no better yields
appear to be obtained. Berkhout estimated 17 lbs. of
dry rubber per acre in the eighth year from planting,
26 lbs. in the tenth year, and 70 lbs. an acre in the
twentieth year. Such yields render the trees almost
useless for plantation purposes, and in many parts of
the Dutch East Indies where they have been planted
they are already being cut down to make way for
Hevea.

Another disadvantage of *Ficus elastica* is the very rapid spontaneous coagulation of the latex, which makes it impossible to collect and treat the latex by ordinary methods.

Other species.

The above comprise all the species which have been used at all extensively in plantations proper. The method of replanting in the forests in which the wild rubber plants have been partly destroyed is being adopted in several countries, notably in Brazil and in the Mabira forest of Central Africa. Rubber vines were extensively planted some years ago in the Congo by the orders of the Belgian Government. This method was recently given up in favour of *Funtumia* plantations, and still more recently there has been a tendency to abandon *Funtumia* in favour of *Hevea*. On the whole, it seems probable that in the future the world's supply of rubber will depend more and more upon the plantations of *Hevea*, including those established or about to be established in Brazil.

CHAPTER X

THE chemical composition and behaviour of india-rubber are among the most difficult problems with which the organic chemist is confronted. The nature of these problems can only be indicated here in the barest outline. For further information the works of Weber and of Schidrowitz should be consulted. The latter gives references to the original chemical papers down to the end of 1910.

Technically pure rubber consists of crude commercial rubber which has been thoroughly washed and dried *in vacuo*. Much of the rubber produced on plantations is therefore already practically in this condition. The difficulties which stand in the way of the further study of this substance may be stated in the words of Weber :—" These difficulties are physical rather than chemical, that is to say they do not so much consist in the functional complexity of india-rubber as in the circumstance that these molecules are only known with the colloidal state superimposed upon them."

Properties of Colloids.

A colloid may be defined as a substance which forms a jelly-like solution, incapable of passing through a membrane of parchment or similar material. Colloids are thus contrasted with crystalloids which are able in solution to pass through colloidal membranes. A colloid does not crystallize, and has no definite melting point; in fact, the physical changes induced in it by heat or by solvents are perfectly continuous so long as the chemical molecule of the substance remains intact. The isolation of a particular colloid from a mixture containing other colloids offers great technical difficulties, and such a substance is exceedingly difficult to obtain in a chemically pure condition. All the ordinary chemical tests of identity are baffled by the consistency of such a material. The colloidal condition is believed to be generally associated with the existence of a very large chemical molecule.

Composition of technically pure Rubber.

Technically pure rubber is by no means pure in the chemical sense. The chief impurities present are various resinous and allied bodies, which can be removed more or less completely by prolonged extraction with boiling acetone. The amount of the resins and oily bodies extracted in this way varies from 1 per cent. upwards, according to the origin of the

particular sample. Proteins and other nitrogenous
bodies are also invariably present in varying amounts,
and there is always a certain amount of ash remaining
on ignition. After the various impurities have been
determined as accurately as possible, the weight of
these is deducted from the original total weight, and
the residue is regarded as representing the amount of
pure caoutchouc present. The approximate composition of biscuit rubber prepared from the latex of old
Hevea trees at Henaratgoda was as follows:—

Resin, etc.	...	...	...	2 per cent.
Proteins, etc.	...	...	...	I „
Ash	...	...	...	0·3 „
Rubber	...	...	...	96 „

Physical Properties.

In the highest attainable state of chemical purity,
india-rubber is a practically colourless and odourless
substance, possessing a specific gravity of approximately
0·911. It is insoluble in water, but is capable of
absorbing about 25 per cent. of its own weight of
water on prolonged immersion. Rubber is a bad
conductor both of heat and of electricity. Some of
its most important technical uses are dependent upon
the last named property. In so-called rubber solvents
the rubber swells into a jelly which finally " runs "
when a sufficient amount of the solvent is added. In
this way about 15 parts of rubber are soluble in 100
parts of benzene, and smaller amounts in turpentine,

petrol, chloroform, carbon bisulphide and other solvents. The rubber is more easily taken up by the solvent after it has been mechanically kneaded in the rolling or masticating machines, than in the untreated condition. Weber considered that all these solutions should more properly be regarded as solutions of the so-called solvent in the india-rubber.

The mechanical strength of rubber which has once been dissolved is greatly inferior to that of the crude rubber. For this reason cut sheet rubber is superior for many purposes to sheet which has been " spread " from a solution. This fact renders it doubtful whether synthetic rubber can ever be made equal in physical properties to natural rubber. The mechanical kneading to which raw rubber is subjected in the course of manufacture has also a marked effect upon its physical properties.

The value of rubber for many purposes depends largely upon its chemical indifference, i.e. upon its want of power to enter into combination with many other chemical substances. Raw rubber undergoes slow oxidation on exposure to the air, and it enters rapidly into combination with ozone. Dilute acids and alkalies have little or no effect upon it, but it is rapidly destroyed by the action of strong sulphuric or nitric acid, or by exposure to the action of chlorine, bromine or iodine. The only compounds of rubber which have been at all closely studied are those with sulphur, with the halogens and with ozone.

On cooling raw rubber to the freezing point of water the substance becomes hard and brittle, but recovers its ordinary properties when restored to the normal temperature. At warm summer temperatures the elasticity of the rubber is increased, but on further raising the temperature the rubber becomes softer and more extensible. If the heating is not carried too far, the normal properties of the rubber are recovered on cooling. Heated still further the rubber becomes soft and sticky, becoming actually liquid at some point above 200°C. If cooled again from such a temperature the normal elasticity of the rubber is no longer recovered. At still higher temperatures the rubber undergoes dry distillation with destruction of the molecule.

Destructive Distillation.—Synthesis.

The destructive distillation of india-rubber gives rise to isoprene C_5H_8, and to other bodies of the formula $(C_5H_8)_n$. Isoprene is also the most important step in the synthesis of rubber, for which a number of commercial processes have recently been patented. Tilden, many years ago, obtained true rubber from isoprene which had been prepared by passing the vapour of turpentine through heated tubes. The rubber appeared spontaneously in the isoprene which had been kept for some years in a closed bottle, and was also produced more rapidly in small quantities by the action of hydrochloric acid. The last named method of preparation confirmed a previous discovery by Bouchardet.

The production of rubber from isoprene is a process of *polymerisation*, that is to say the union of a number of molecules each possessing the same *empyrical formula*, i.e. containing the same relative number of carbon and hydrogen atoms. Analysis has shown that rubber contains carbon and hydrogen in the same proportions as exist in isoprene and turpentine. A study of the compounds of rubber, particularly those with ozone, has led Harries to represent the polymerisation of isoprene in the following manner:

$$
\begin{array}{ccc}
\mathrm{H_2{=}C} & \mathrm{C{=}H_2} \\
\parallel & \parallel \\
\mathrm{H{-}C} & \mathrm{C{-}CH_3} \\
\mid & \mid \\
\mathrm{CH_3{-}C} & \mathrm{C{-}H} \\
\parallel & \parallel \\
\mathrm{H_2{=}C} & \mathrm{C{=}H_2}
\end{array}
\quad = \quad
\begin{array}{c}
\mathrm{H_2} \qquad\qquad \mathrm{H_2} \\
\mathrm{C{-}C} \\
\mathrm{H{-}C} \qquad \mathrm{C{-}CH_3} \\
\parallel \qquad\qquad \parallel \\
\mathrm{C} \qquad\qquad \mathrm{C{-}H} \\
\mathrm{CH_3} \qquad \mathrm{C{-}C} \\
\mathrm{H_2} \qquad\qquad \mathrm{H_2}
\end{array}
$$

The actual molecule of rubber is probably a polymer of the substance whose constitution is represented above, and may consist of a number of similar eight-ring molecules linked together. The empyrical formula of rubber is thus $C_{10}H_{16}$, which is identical with those of turpentine and gutta-percha. The molecule of the latter is probably even more complex than that of rubber, but may possibly be a higher polymer of the same series. It should be observed that the account here given of the constitution of the rubber molecule only represents one of the most recent theories. The problem of the structure of this molecule cannot yet be regarded as finally settled.

In some of the modern processes which have been proposed for the manufacture of synthetic rubber on a commercial scale, isoprene is prepared from fusel oil, obtained from starch by special methods of fermentation; and polymerisation is effected by the action of metallic sodium. Various other substances have also been synthesised, which, though differing in composition, appear to possess a similar molecular structure to indiarubber. The physical properties of some of these bodies are closely similar to those of rubber, and their use has been proposed commercially for similar purposes. So far, none of these substances have been placed upon the market in appreciable quantities, and the reports of the cheap production of synthetic rubber have not carried with them a fall in the value of plantation shares.

Vulcanisation.

The molecule of rubber is unsaturated, and is able to enter into direct combination with different elements. This fact is expressed in its constitutional formula by the presence of double links between two pairs of carbon atoms in the ring. Among the addition products of rubber, those with sulphur possess the greatest technical importance. Rubber combined with sulphur is said to be vulcanised. We have already referred in Chapter I to the remarkable properties of resistance and durability possessed by vulcanised rubber.

Rubber is vulcanised by two entirely different processes. In hot vulcanisation the rubber is directly combined with sulphur under the influence of heat. In cold vulcanisation, rubber is combined, without heating, with chloride of sulphur dissolved in benzene or other solvent. The molecule of rubber vulcanised by the second method contains chlorine as well as sulphur.

Vulcanisation with hot sulphur was discovered by Nelson Goodyear in 1839. At the melting point of sulphur there is little or no action. Combination begins at about 120°C., and the temperatures used in practice range from 125°C. upwards. According to the amount of sulphur employed, the time of action and the temperature, the final product ranges in character from soft elastic to hard vulcanite.

Weber considered that the process of vulcanisation was one of definite chemical combination. He believed that a series of sulphides of rubber were formed ranging from $(C_{10}H_{16})_{20}S_2$ to the final product vulcanite, to which he attributed the formula $C_{10}H_{16}S_2$. This opinion, which has been subjected to severe criticism, appears to be confirmed by the results of the most recent researches. The process is a reversible one, and the vulcanised rubber apparently always contains a certain amount of uncombined rubber and a certain amount of free sulphur. As a result, goods which have only been partially vulcanised become more fully vulcanised on keeping. It is therefore the practice to vulcanise the majority of goods rather less completely than is finally required.

Some recent writers still regard the process of vulcanisation as one of adsorption. That is to say they look upon the vulcanised rubber as existing in a condition of physical mixture rather than in one of true chemical combination. As previously stated, the whole problem is one of extreme difficulty, and we are perhaps still far from a final solution.

The physical condition of the rubber has a marked influence upon the final product, and the same sample of raw rubber, if differently worked before vulcanisation, requires to be vulcanised with a different quantity of sulphur and to a different extent in order to produce the same final result. After long continued kneading and mastication more sulphur is required in order to bring the vulcanised rubber to the same condition of physical consistency.

The process of cold vulcanisation with sulphur monochloride was discovered by Parkes in 1848. The action is exceedingly vigorous. Dilute solutions are therefore employed, and the period of contact is short. The solvent almost universally employed is carbon bisulphide.

The action of sulphur monochloride on rubber, like that of sulphur, is an addition and not a substitution process. This is proved by the fact that no sulphuretted hydrogen is emitted during the process of combination. The action probably gives rise to a series of addition-products parallel with those arising during simple vulcanisation with sulphur. The series would then begin with $(C_{10}H_{16})_{20}S_2Cl_2$ and end with $C_{10}H_{16}S_2Cl_2$. The fact that the addition of large amounts of the pure solvent leads to the extraction of S_2Cl_2 from the vulcanised rubber is equally well explained on the hypothesis of a reversible chemical action as it is on the adsorption hypothesis.

CHAPTER XI

THE principal processes employed in a rubber
factory begin with the washing and drying of the crude
commercial samples, which require thorough cleansing
in order that technically pure rubber may be obtained.
From this point two series of operations diverge,
which may be distinguished as wet and dry processes
respectively. In the former the rubber is dissolved
in naphtha or benzene, and is then deposited from the
solution in moulds or on cloth. The articles so formed
are subsequently vulcanised, often by the cold process,
i.e. the action of sulphur monochloride. The bulk of
the larger and more solid goods are manufactured by
a dry process. In this the rubber is first masticated
or kneaded, in order to bring it into a suitable condition
for the next operation, that of mixing, in which various
filling and diluting materials are intimately distributed
through the substance of the rubber, together with the
sulphur required for vulcanisation. The mixed rubber
is then moulded or forced or built up with layers of

canvas into the various articles which are being made in the factory. Vulcanisation is subsequently carried out by steam heat in moulds or presses or closed chambers. Certain other articles again are cut out of rubber already vulcanised.

Washing.

It has already been pointed out that in dealing with the pure samples of rubber now produced on estates, the preliminary cleansing operations, necessary in the case of all ordinary commercial samples, will probably be omitted from the series of manufacturing processes in the rubber factories of the future. The washing machines used in factories are generally similar to those employed upon estates, but are usually larger and more powerful. They are generally driven from shafting through powerful clutches, and special arrangements are often introduced for throwing the machinery automatically out of gear in case of accident, for example when large stones or other bodies liable to cause damage are encountered. Prior to the actual process of washing, large commercial blocks are softened by soaking in hot water, and are sliced into smaller pieces by special machines, either before or after the softening process. A series of washing machines is usually employed, and the rollers through which the rubber first passes have coarser grooves and are set further apart than the later members of the series. The arrangement thus resembles to some extent the series of rollers in a modern flour mill, although the processes of rubber washing and flour milling have little other resemblance. By passing repeatedly through the washing rollers under a stream of water, the impurities present in most kinds of commercial rubber are gradually washed away, and the rubber finally emerges in the form of a thin crêpe convenient for drying. The rubber itself undergoes much tearing and stretching during the process, and it is important that this mechanical action should not be carried further than is absolutely necessary in order to ensure a sufficient degree of purity, since the "nerve" of the rubber is largely destroyed by the forcible treatment involved.

Drying.

The so-called "nerve" lost in washing is partly recovered during the slow process of drying, which is effected by hanging the strips of rubber in large chambers exposed to a current of dry air. The drying of the washed rubber takes place more readily than in the case of rubber freshly prepared from latex, since the moisture is not so closely incorporated in the substance of the rubber. On the other hand many commercial rubbers, for example hard Para, contain a considerable proportion of the original moisture derived from the latex. When this moisture in addition to the water introduced during washing has to be removed, the drying process takes considerably longer.

Vacuum driers are also under trial in certain factories, but it is said that rubber dried in this way does not recover its "nerve" so completely as it would if the slower process of air drying were adopted. Drying machines of this type are, however, frequently employed for desiccating the materials used for mixing with the rubber, thorough dryness being essential in the latter process.

Mastication and Mixing.

Mastication is sometimes carried out in a machine resembling a powerful churn or sausage machine. For most purposes rollers are employed similar to or identical with the mixing rollers. Mixing is always preceded by a masticating process, introduced in order to soften the rubber. The dried rubber is passed repeatedly between a pair of hollow rollers, which can either be heated internally by steam, or cooled by passing cold water through the cavity. The rollers revolve at different rates, and the distance between them is adjustable. The rollers are gradually brought closer together until the soft rubber adheres to the slower moving roller and passes round with it, being subjected to a kneading process as it passes the second roller, which revolves more rapidly. At the beginning of the process the rollers are warmed, but later on friction may give rise to so much heat that it is necessary to pass cold water into the interior of the rollers. The whole process thus calls for considerable skill and judgment on the part of the operator in regulating both the temperature and the action of the machine.

After undergoing a certain amount of mastication the rubber is judged to be ready for mixing. In the mixing process, the sulphur required for vulcanisation is incorporated with the rubber, together with other substances known as fillers. The rubber is passed repeatedly between the rollers and the sulphur, and filling materials are gradually sprinkled over it in the form of a fine powder. The process of rolling is then continued until the fillers are evenly distributed through the rubber, and the whole substance has become nearly homogeneous.

Sulphur is introduced either in the form of flowers of sulphur or as precipitated sulphur. Other substances are used in the mixing for several definite purposes. Thus various sulphides aid in the process of vulcanisation. These and other chemicals are also introduced in order to impart definite qualities of toughness and durability to the finished product. Among the most important of these substances are zinc oxide, magnesia, antimony sulphide and litharge (lead sulphide). Some of these chemicals impart characteristic colours to the rubber, whilst a further series of materials is used simply for colouring. Thus red rubber may contain vermilion, red lead or antimony sulphide ; whilst white or grey rubbers usually contain zinc oxide, and black rubber contains either litharge or some form of carbon black. Finally, there are the filling materials which are introduced in order to reduce the specific gravity of the rubber or to lower the cost. Rubber is sold by weight, and the materials used for toughening have mostly high specific gravities. Substances of low specific weight are therefore introduced in order to counteract the effect of the heavy bodies. For this purpose whiting and French chalk are largely employed. For cheapening, a large variety of substances is used, foremost among them being old vulcanised rubber in a more or less reclaimed condition. Fatty oils undergo a kind of vulcanisation with sulphur, and are used as cheapeners in combination with reclaimed rubber. With low grades of rubber such substances as leather waste and sawdust may even be incorporated.

The further treatment of the mixed rubber varies according to the kind of article which is to be manufactured. Very commonly it is treated in one of three ways. It is either pressed into blocks, or rolled into sheets, or moulded in what is known as a forcing machine.

Preparation of Sheet Rubber.

A large proportion of rubber articles are manufactured from sheet rubber. The highest class of sheet rubber is cut from block, but this method is less used than formerly, on account of the increased perfection of other methods. For making " cut sheet," the rubber, after masticating and mixing, is pressed into a solid block. The block is then frozen hard, an operation which takes a long time owing to the low conductivity of rubber for heat. After freezing, the block is cut into sheets by one of two methods. In the older method the block of rubber was sliced horizontally, being raised, after the removal of each sheet, by an amount equal to the thickness of the sheet. In the second method a cylindrical block is made to rotate against the knife blade, which thus slices off a continuous sheet. The knife is kept wet during the process, and special methods of sharpening and setting have to be adopted.

Raised sheet is prepared by spreading rubber in solution over cloth, in a very thin film. A series of such films are spread over the cloth, the solvent being allowed to evaporate between each spreading. Finally the cloth is detached from the sheet of rubber thus produced. A similar process is employed in water-proofing, in which a thin layer of rubber is permanently attached to the surface of the cloth. Sometimes two layers of cloth are united by a thin layer of rubber placed between them. It was for waterproofing that rubber in solution was first employed. Sheet rubber prepared on cloth in this way carries the grain of the cloth upon its surface. Sheet having a perfectly smooth surface can also be prepared from solution by spreading the dissolved rubber over the surface of a plate of glass.

Calendering.

At the present day by far the largest proportion of sheet rubber is prepared in machines known as calenders. Indeed, the calenders are among the most important machines in modern rubber factories. A calender consists of a series of heavy rollers revolving alternately in opposite directions. A common type has three rollers arranged one above the other. The rubber mixture is fed between the upper and middle rollers. The sheet thus formed, varying in thickness according to the distance at which the rollers are set, adheres to the middle roller and passes back between the middle and lower rollers. For packing the sheet thus obtained, the following method is adopted. A sheet of cloth is unwound from a drum on one side of the machine. The cloth passes between the middle and lower rollers, and is rolled up together with the rubber sheet on the opposite side of the machine. If it is desired to make the rubber adhere closely to the cloth, the middle and lower rollers are made to revolve at different rates of speed, and the friction thus caused produces the desired adhesion. The rubber is then said to be "frictioned" on to the cloth. Frictioned sheet is used in the manufacture of a large variety of articles in which alternate layers of cloth and rubber have to be incorporated.

Manufacture of Various Articles.

A large variety of articles are made directly from sheet rubber. The highest class of cut sheet rubber is used in the manufacture of such articles as tobacco pouches. Permanent joints can be made by simply pressing together the cleanly cut edges of the sheet. In order to prevent adhesion at other points, French chalk is dusted over the surface. If lower grade sheet is used, it may be necessary to moisten the edges with solvent or with rubber solution in order to make a permanent joint.

Block rubber is reconstructed from sheet for the manufacture of such articles as railway buffers. The sheet is rolled up into a cylinder and firmly pressed together. Greater strength is thus obtained than by simply blocking the mixed rubber. The nature of the increased strength may be understood from the analogy of heavy guns wound from wire, which are much stronger than the old fashioned guns cast in one piece.

Large tubes, such as the inner tubes for motor tyres, are made by rolling the sheet rubber round an internal mould known as a *mandrel.* The edges of the sheet are joined together with solution. The ends of the tube are cemented up after vulcanisation and removal from the mandrel. The removal of the tube is effected by the aid of compressed air. Smaller tubes of a simple kind, and such articles as solid rubber tyres, are made by "squirting" the rubber mixture through a die by means of a forcing machine. Such a machine consists of a large screw working inside a cylinder, and so forcing the rubber forward after it has been fed in through a hopper at the top of the machine. In making tubes a rod or mandrel is made to pass through the centre of the screw and of the die, and the rubber is forced round the mandrel, which is moved forward at a suitable rate. The mandrel may be first coated with canvas and the rubber forced over it. Another layer of canvas can then be applied, and the forcing process repeated, until a large tube has been built up of alternate layers of canvas and rubber. Other tubes may be built up by hand on a mandrel with alternate layers of sheet rubber and cloth. The covers of electric cables are similarly constructed, or they may be wound with thin ribbons of rubber in a special machine. A large variety of other articles are prepared by moulding or pressing the rubber whilst in the plastic condition.

In making india-rubber balls and other hollow objects, a small blob of india-rubber containing no sulphur is placed against one side internally. When the ball cools after vulcanisation the walls collapse. The ball is then inflated through a fine-pointed syringe pushed through the lump of soft rubber. When the syringe is withdrawn the hole is closed airtight by the unvulcanised rubber, and the inflated condition is thus preserved.

Various other simple articles are cut from rubber which has already been vulcanised. Rubber rings, for example, are sliced off the ends of vulcanised tubes.

Rubber Solution.

A large variety of objects is manufactured by dipping prepared fabrics in rubber solution. The solvents chiefly employed are benzene and mineral naphtha. The rubber is rendered more readily soluble by undergoing previous mastication. Rubber shoes and galoshes are dipped by dozens at a time in tanks of solution and afterwards allowed to dry. Teats for feeding bottles are made by dipping glass moulds repeatedly in clear solution, with intervals for drying. Some of the smaller articles prepared in this way are vulcanised by the cold process, but for larger objects the sulphur required for vulcanisation is incorporated in the solution. For repairing motor tyres a quick-vulcanising mixture is employed, in order that the rest of the tyre may not be damaged by the heat required for vulcanising the mended area.

Vulcanisation.

H. C. Pearson has pointed out that the natives of some parts of the Amazon districts are accustomed to mix sulphur with the latex of *Hevea* before employing it for waterproofing. True combination between the rubber and the sulphur, however, apparently only occurs at a high temperature. We have already distinguished in the last chapter between hot and cold vulcanisation. Under hot vulcanisation again two processes may be separately considered. Of these, the more important is the dry process originally patented by Goodyear. In this process the proper amount of sulphur is incorporated with the rubber in the mixing rollers before the articles are made up, and vulcanisation is effected by heating either directly in steam under pressure or in steam-jacketed chambers, or in hydraulic presses heated by steam in various ways.

Hancock's wet process, on the other hand, is chiefly of historical interest. In this process articles made of sheet rubber, without any previous admixture of sulphur, are immersed in molten sulphur for a certain time and at a certain temperature.

Finally, in the cold process, patented by Parkes in 1848, the articles are immersed in a solution of chloride of sulphur, in carbon bisulphide, or in benzene. This process is exceedingly rapid, and can only be applied to articles of very thin sheet.

The Dry Process.

The chambers used for vulcanising by the dry process are often of very large size. They may take the form of great iron tunnels, into which the articles to be vulcanised are run on rails. Hose pipes are often vulcanised in lengths of 60 feet or more. The rubber becomes soft and plastic at the temperature employed, and must therefore be supported in order to preserve its shape. Powdered French chalk is largely employed as a bed for the rubber articles to lie on. Closed objects such as balls, india-rubber dolls and other toys, are vulcanised in moulds, and before the cavity of the rubber is closed up some substance is introduced which will volatilise at the temperature of vulcanisation, and so press the object firmly against the mould. The outer covers of motor tyres are vulcanised in heavy presses, in order that their substance may be firmly compacted during the process. The most complicated of all vulcanising machines is the autoclave press, which consists of a hydraulic press completely enclosed in a steam pressure chamber. In this way the difficulties of obtaining an even temperature in a press heated by steam pipes are overcome. Vulcanisation by steam is generally carried out at a pressure of three to four atmospheres (45 to 60 lbs. per square inch) corresponding to a temperature of 134°C. to 144°C. The temperature and pressure are first raised gradually, and afterwards kept high for three or four hours. Self-recording thermometers and pressure gauges are fitted to the apparatus in order that the process may be kept under complete control. The time required for vulcanisation varies according to the source of the rubber, being shortest in the case of *Hevea* rubber. The previous treatment of the rubber also affects the process. Thick articles naturally require to be heated for a longer period than thin ones. The proportion of sulphur added to the rubber varies according to the nature of the articles to be manufactured. For ordinary goods the quantity is about 7 to 10 per cent. of the amount of rubber.

Bath Process.

In Hancock's bath process, which is comparatively little used, small objects are immersed in molten sulphur for a period of two or three hours at a temperature of 130°C. to 135°C. Test pieces of rubber, of similar thickness and composition to the articles to be vulcanised, are placed in the same bath and removed at intervals. From the appearance of these the operator is able to judge when the process of vulcanisation is complete.

Cold Process.

The cold process, discovered by Parkes, can only be used for vulcanising very thin sheet. This is due to the extreme rapidity of the process, which is such that if it were applied to thick rubber, the surface of the object would be converted into vulcanite before the interior was properly vulcanised. The reagent employed is *sulphur monochloride*, prepared by passing dry chlorine over heated sulphur. This substance is very active chemically, and is decomposed by water. As a solvent, carbon bisulphide is almost universally adopted. Carbon bisulphide is exceedingly poisonous and inflammable, but no other solvent has been found to give equally good results. The strength of the solution used is generally about $2\frac{1}{2}$ per cent. The sheet rubber, which must first be very thoroughly dried, is immersed in the solution for three minutes or less, according to its thickness.

Whatever the method of vulcanisation employed, there is always a certain amount of after effect, owing to the prolonged slow action of the residual rubber. Articles are therefore almost always rather less perfectly vulcanised than their final condition requires. The goods afterwards improve to some extent by keeping. In the case of articles in which a high degree of permanent elasticity is required, the excess of sulphur must be removed by boiling in caustic soda, followed by a thorough washing in water in order to remove the alkali. This process is necessary, for example, in the manufacture of elastic thread, which is cut from vulcanised spread sheet and afterwards freed from sulphur in the manner described.

Reclaimed Rubber.

During recent years the scarcity and high price of fresh rubber has led to the use of enormous quantities of old rubber, reclaimed in various ways, either in combination with fresh rubber or alone. Old rubber may be simply ground to powder and employed as a filling material with fresh rubber, but the bulk of the reclaimed rubber used in manufacture goes through a series of complicated processes. The discarded rubber goods used for making reclaim, whether they be old rubber shoes or old motor tyres, usually contain a considerable quantity of cotton fibre. In order to remove this fibre the rubber is boiled with dilute sulphuric acid, and the cotton thus disintegrated can then be got rid of by grinding, combined with the use of an air blast. The acid process is generally followed by an alkali process, in which the free sulphur is removed by boiling with caustic soda. The residual rubber is then heated with resin-oil, and can afterwards be manipulated more or less like ordinary unvulcanised rubber. The reclaimed rubber contains all the mineral substances originally added, so that little further mixing is required if a similar class of goods is to be remanufactured.

In addition to the methods described above, it is claimed that processes have already been perfected by which the sulphur of vulcanisation can be more or less completely removed from the old rubber. Schidrowitz states that other processes have been recently introduced by which "particles of vulcanised rubber in the shape of dust or flakes can by pressure and heat be moulded to a homogeneous mass, which, on cooling, is to all intents and purposes indistinguishable from an ordinary moulded article."

Vulcanite.

The manufacture of vulcanite resembles in principle that of ordinary rubber goods at all stages, except that a larger proportion of sulphur is added in the mixing and that vulcanisation is carried out for a longer period and at a higher temperature. As we proceed to stages of vulcanisation beyond that for ordinary hard vulcanised rubber, we pass through the tough and springy condition of the substance employed as artificial whalebone, and finally arrive at the hard and brittle state of the vulcanite or ebonite employed for making fountain pens and the mouthpieces of pipes—to name only two of a large variety of uses. In making whalebone-substitute the amount of sulphur added to the rubber is from 12 to 14 per cent.; for the harder ebonite from 24 to 35 per cent. is required. Vulcanisation is carried out either for a prolonged period— 8 to 12 hours—at a comparatively low temperature— about 135°C., or for a shorter time at a higher temperature.

The Testing of Rubber Goods.

A number of ingenious devices are employed in determining the quality of rubber goods as well as that of raw rubber. The principal tests to which manufactured rubber is subjected are those for abrasion and for stretching. In abrasion tests the sample is brought in contact with a rotating wheel of definite composition for a certain time and with a certain pressure, and the loss in weight of the rubber is afterwards determined. For stretching tests special machines are used in order to cut perfect rings from sheet of uniform thickness. The ring is tested by passing it round two rollers, which rotate at an even speed whilst they are gradually moved further and further apart by a stretching force, the amount of which is recorded upon a dial. The force required in order to produce a certain elongation, or to break the ring, may thus be determined.

CONCLUSION.

We have now followed the fortunes of rubber from the wild territories of the Amazon to the plantation, and from the plantation to the factory. Here we may take leave of it, conscious that the last word on the subject is far from being said. In fact, we shall probably be safe in asserting that the future history of the rubber industry will be at least as interesting and eventful as its past history. At every stage of its career, rubber still presents problems which the planter and the manufacturer must join hands in solving, with the help of their scientific advisers.

INDEX

Acre, 21

Acreage under rubber, 12

Africa, Central and East, plantation industry in, 11

West, 33

African rubber, 30 *et seq.*
largely exterminated, 30
new plantations of, 32

Agricultural machinery, 99

Alstonia, 36
plumosa, 36

Amazon region, 17
foodstuffs in, 23
rubber industry, economic aspects of, 21

Amazonas, 21

American species of rubber, 17 *et seq.*

Angle of cut, 145

Angola, 33
rubber planting in, 11

Animal pests of *Hevea*, 179

Asiatic rubbers, 34

Assam, 34, 208

Balls, india-rubber, 229

Bamber, M. Kelway, 56
pricking method, 134

Bark, effects of wounding the, 47
renewal of, 49
ringing the, 47, 48

Basal system of tapping, 139

Bath process, Hancock's, 233

Biffen, centrifugal method of separation, 159

Biscuit rubber, 154
approximate composition of, 212

Block rubber, 164, 167

Bordeaux mixture, for Pink disease, 190

Boring insects, 52

Borneo, 36

Botanical sources of rubber, 16 *et seq.*

Botryodiplodia theobromae, 191

Bouchardet, discovery by, 215

Bowman-Northway pricker, 132

Brain, Lewton, estimate by, 13

Brazil, collection of rubber in, 19
labour in, 23
replanting wild rubber in, 209
transport in, 23

Brazilian Government, encourages rubber planting, 12
legislation by, 22
premiums offered by, 23
special concessions by, 23

British India, wild rubbers from, 5

British West Africa, *Funtumia* in, 207

Brown root disease, 186

Burma, *Ficus elastica* in, 34
Hevea plants sent to, 7

Burrs, 192

Cacao, as intercrop, 114

Calendering, 227

Cambium, 41, 44, 48

Cambodia, 35

Canker, 187
of *Hevea* fruits, 189
remedy of, 188

Capital in rubber, 15

Carpodinus lanceolatus, 33
Castilloa, 8, 26, 197 *et seq.*
  bark of, 27
  in Central America, 198
  in Ceylon, 199
  in Jamaica, 200
  in Mexico, 198, 200
  in New Guinea, 200
  in West Indies, 200
  Markhamiana, 27
  tapping of, 28, 199
  wound response in, 59
  yield from, 198
Caucho rubber, 26
Ceara, 24
  at Peradeniya, 204
  coagulation of, 203
  difficulties in tapping, 25
  growth of, 201
  in Brazil, 201
  in Ceylon, 200
  in Hawaii, 203
  in Nyassaland, 204
  in Zanzibar, 201
  seeds of, 25
" Ceara scrap," 25
  tapping of, 24, 202
  tapping difficulties, 200
  yields from, 204
Centrifugal method, 159
Ceylon, *Castilloa* in, 199
  draining in, 104
  Hevea in, 7 *et seq.,* 95
  labour in, 123, 124
  land tenure in, 97
  nodules in, 193
  plantation industry in, 10
  plantation yields in, 152
  planting distances in, 106
  rate of growth in, 111
  tools used in, 126
  transport in, 97
  wages in, 125
Chemistry of rubber, 210 *et seq.*
Christy, C., on *Funtumia,* 205
Clearing, 98
Clitandra henriquesiana, 33
Coagulation, 54, 153, 157 *et seq.*
Cochin China, 35

Cockerill, electrolytic method of
  separation, 159
Coffee, as intercrop, 114
Coffee leaf fungus (*Hemileia vasta-*
  trix), 177
Cold process, Parkes', 233
Collection of rubber in Brazil,
  19
Collins, James, 5
  brings first seeds of *Hevea,* 6
Colloids, 211
Columbus, 3
Condamine, C. M. de la, 5
Congo, 33
  rubber vines planted in, 209
  wild rubbers from, 5
Cortex, 40
Cotton, as intercrop, 114
Cover crops, 112, 113
Crêpe, 162, 163
  pale, 164
Crêping, 164
Cross, 5
  brings *Castilloa* seeds, 6
  introduces Ceara rubber to
    Kew, 6
Crotalaria, 113
Cultivation, 115
Cut, angle of, 145
  direction of, 146
Cuts, distance between, 146

Deep forking, 115
Die-back, 191
  remedy for, 192
Diseases of *Hevea* roots, 183
Draining, 103
  in Ceylon, 104
  in Malaya, 104
  on hillsides, 105
Drought, 112
Drying, 162 *et seq.,* 222
Dry rubber, analysis of, 85
  crop in lbs., 73
  yield of, 77
  yields of, at Henaratgoda,
    62
  yields per acre, 64
Dust mulch, 113

Dutch East Indies, rubber cultivation in, 11
Dyera, 36

East and West Indies, suggested by
Hancock for rubber produc-
tion, 5
Economic aspects of Amazon rubber
industry, 21
Enzyme, 55
Epidemics, 177
Erythrina lithosperma, 119
Estate expenses, 125
factory work on, 153
general sanitation of, 194
routine of, 129
Estimate, for planting 500 acres, 127
Estrade, 20
Excision methods of tapping, 59,
136
Exhibition, World's First Rubber, at
Peradeniya, 10
Exhibitions, in Rio de Janeiro, 23
Exports of rubber from East, 13
from Malaya, 13

Factory, cleanliness in, 156
lighting of estate, 155
machinery for estate, 155
site of, 154
work, on estate, 153
Federated Malay States, area of
rubber plantations in, 11
draining in, 104
plantation industry in, 10, 11
Ficus elastica, 34
in Assam, 208
in Java, 208
Sir Daniel Morris on, 208
tapping of, 35
yields from, 208
Ficus Vogelii, 33
Fiji, 36
Fitting, on food supplies of tree, 49
on tapping, 48
Fomes semitostus, 183
Forsterionia floribunda, 29
gracilis, 29
Fungus diseases, 182

*Funtumia elastica (Kickxia afri-
cana)*, 27, 30, 197, 205 *et seq.*
coagulation of, 207
laticiferous system of, 43
pests affecting, 207
planting of, 205
seeds of, 205
tapping of, 30, 206
wound response in, 59

Galoshes, 230
Gloeosporium alborubrum, 191
Gold Coast, wild rubber exported
from, 5
Goodyear, Nelson, discovers vul-
canisation, 1, 217
Growth, rate of, in Ceylon, 111
in Malaya, 111
Guayule rubber (*Parthenium ar-
gentatum*), 29
processes of, 30
Guttering, 150

Half-herring-bone system, 137
Half-spiral system, 139
Hancock, Thomas, bath process,
233
employs rubber for waterproofing, 1
suggests cultivation of rubber,
5
Hancornia speciosa, 28
Harries, on isoprene, 215
Harvesting operations, 128
Hawaii, Ceara in, 203
Hayti, 3
Henaratgoda, 7
experiments at, by the author,
56
yield from one tree, 66–67
Herring-bone system of tapping,
133, 139
Hevea brasiliensis, 4, 16, 17 *et seq.*
bark, minute anatomy of, 45
canker of fruits, 189
choice of situation and soil, 93
early experiments in Ceylon, 9
first seeds brought from America, 6

Hevea brasiliensis, continued
  flowers of, 19
  fungus diseases of, 182
  gross structure of bark, 44
  in Ceylon, 8
  irrigation in cultivation of, 105
  latex, coagulation of, 54
  latex, composition of, 53, 54
  latex vessels in, 43, 45
  leaf-fall, 19
  pests of, 176
  planting operations, 93
  plants arrive in Ceylon, 7
  plants sent to Burma, Java,
    Singapore and W. Indies, 7
  plants sent to Mauritius, W.
    Africa and Fiji, 7
  repeated tapping of, 91
  seeds of, 19
  seeds, 122
  tapping experiments in, 56
  section of bark, 46
  species of, 17
  Spruceana, plants sent to Cey-
    lon, 7
  wound response in, 59
  yields of rubber from one tree,
    9
Hill, H. C., on *Ficus elastica*, 208
Holing and planting, 109
Hooker, Sir Joseph, 5
Hose-pipes, 231
Hymenochaete noxia, 184, 186

Incision, methods of, 59
India, rubber planting in, 11
India-rubber balls, 229
  chemistry of, 210
Indigo, as intercrop, 114
Indigofera, 113
Insects, pests of *Hevea*, 180 *et seq.*
Intercrops, 108, 113 *et seq.*
  cacao, 114
  coffee, 114
  cotton, 114
  in Ceylon, 113
  in Sumatra and Java, 114
  indigo, 114
  tea, 114

Irrigation, 103
Isoprene, 214 *et seq.*

Java, 7, 35
  Ficus elastica in, 208
  labour in, 122
Jelutong rubber, 36
  export of, 37

Kamarun, *Funtumia* in, 207
Kerckhove, Van den, estimate by, 12
Kickxia africana, see *Funtumia
  elastica*

Labour, in Brazil, 21
  in Ceylon, 123, 124
  in Java, 122
  in Malaya, 123, 124
  in S. India, 123
  system of advances, 125
Lagos silk rubber (*Funtumia elastica*), 30
Lagos, wild rubber exported from,
  5
Land tenure, in Ceylon, 97
  in other countries, 97
Landolphia, climbing rubber, 17, 32
  Dawei, 33
  florida, 33
  Heudelotii, 33
  Kirkii, 33
  owariensis, 33
Latex, 38
  bulk extracted in a year, 91
  cleanliness in dealing with, 151
  effect of tapping on, 82
  effect of water supply on, 88
  manufacture of, 69
  movement of, 70
  origin of, 68
  percentage of rubber in, 83–84
  physiology of, 56
  production, physiology of, 38
  protective function of, 52
  seasonal variation in, 71, 84
  seasonal variations in concen-
    tration of, 84–85
  smoking of, 165
  transport of, 156

Latex, *continued*
  tubes, 41, 42
  vessels, 48
  vessels, in *Hevea*, 43, 45
  vessels, in *Manihot*, 43
  vessels, volume of, 67
  yield at certain seasons, 92
Latices, mixing of, 204
Laticiferous system, 43, 44
  in *Funtumia*, 42
  in the seedling, 46, 47
Leucaena glauca, 118
Lining and spacing, 106

Macadam's comb pricker, 133
Machinery, agricultural, 99, 115
Machines, crêping, 162
  macerating, 162
  sheeting, 161
Macintosh, Charles, and Co., 1
Madagascar, 33
Malaya, estimated export of rubber
  from, 13
  labour in, 123, 124
  plantation yields in, 152
  planting distances in, 106
  rate of growth in, 111
  tools used in, 126
  wages in, 125
Mandrel, 228
Mangabeira rubber, 28
Maniçoba rubber, 26
Manihot dichotoma, 26, 204
Manihot Glaziovii (Ceara rubber),
  6, 8, 24, 43, 197, 200 *et seq.*
  formation of latex vessels, 47
  laticiferous system of, 202
  wound response in, 59
Manihot heptaphylla, 26
Manihot piauhyensis, 26
Manufacture of latex, 69
  of rubber goods, 220 *et seq.*
  calendering, 227
  colouring matter, 224
  drying, 222
  mastication and mixing, 223
  sulphur in, 224
  vacuum driers, 223
  washing, 221

Manuring, 115, 116
  at Peradeniya, 116
  green, 117
  phosphates, 116
  potash, 116
Mariaella Dussumieri, 181
Markets, rubber, 169
Markham, Sir Clements, proposes
  plantations, 5
Marking the tree, 140 *et seq.*
Mascarhenasia elastica, 34
Mauritius, 7
Medullary rays, 44, 49
Mexico, Guayule rubber, 30
  rubber planting in, 11
Million acres, produce of, 13
Morris, Sir Daniel, on *Ficus elastica*,
  208
Motor tyres, 228, 232

Nerve, 171
New Guinea, 36
Nigeria, Southern, *Funtumia* in, 207
Nodules, 192
  outbreak in Ceylon, 193
Nurseries, 99
Nyassaland, Ceara in, 204

Oil, from rubber seeds, 121
Omaquas Indians, 4
Overtapping, 78

Packing, 168
Pao di Xirringa (syringe tree), 4, 21
Para, hard, price of, 14
  rubber, 3, 21, and see *Hevea brasiliensis*
  rubber, export of, 4
Parameria glandulifera, 35
Paring, basal system, 139
  full herring-bone, 139
  full spiral, 139
  half-herring-bone, 137
  half-spiral, 139
  physiological effect of, 79
  process, 143
  systems of, 137
Parkes' cold process, 233

Parkes, discovery by, 218
Parkin, experiments by, 60
  incision method employed by, 64
  pricking methods, 131
Parthenium argentatum (Guayule rubber), 29
Pearson, H. C., 230
Peradeniya, Ceara at, 204
Petch, T., on blemishes in rubber, 173
  on dead stumps, 195
  on *fomes semitostus*, 185
  on tackiness, 175
Phloem tubes, 41, 45, 49
Phosphates, in manuring, 116
Physiology of latex production, 38
Phytophthora Faberi, 187
Pink disease (*Corticium salmonicolor*), 190
  Bordeaux mixture for, 190
Plantation industry, rise of, in the East, 10
  rubber, best form of, 170
  yields, 151
  yields, in Ceylon, 152
  yields, in Malaya, 152
Planting distances, in Ceylon, 107
  distances, in Malaya, 106
  hexagonal, 108
  on the square, 108
Polymerisation, 55, 215
Potash, in manuring, 116
Price of rubber, 14
Pricker, Bowman-Northway, 132
Pricking, by Parkin, 131
  by Trimen, 131
  effect of, 58
  methods, 133
  objections to, 132
  Wright on, 130
Production from estates, 12
Pruning, 120
  thumb nail, 120
Pure rubber, 210
  composition of, 211

Quality, 171

Railway buffers, 228
Rainfall at Henaratgoda, 72
  at Peradeniya, 61
  effect on yield, 73
Reclaimed rubber, 234
Renewal of bark, 49
Resiliency of rubber, 171
Response of trees to certain stimuli, 65
Resting periods, 89
Rhizomorphs, 185
Roads, 103
Root rubbers, 33
Roots, diseases of, 183
Rubber, acreage under, 12
  African species of, 30 *et seq.*
  American species of, 17 *et seq.*
  analyses of, 86
  approximate composition of, 212
  Asiatic species of, 34 *et seq.*
  balls, early mention of, 3
  biscuit, 154, 212
  block, 164, 167
  botanical sources of, 16
  capital in, 15
  chemical indifference of, 213
  chemistry of, 210 *et seq.*
  collection of, in Brazil, 19
  defects in, 173
  discovery of, 3
  early uses of, 1
  estate, site of, 96
  exports of, from Brazil, 4
  exports of, from East, 13
  globules, size of, 53
  goods, manufacture of, 220 *et seq.*
  goods, testing of, 236
  loss of weight in washing, 160
  markets, 169
  method of collection, 3
  molecule of, 215
  percentage in latex, 83, 92
  physical properties, 212
  planting, birth of industry, 5
  price of, 14
  quality of, 171
  reclaimed, 233

Rubber, *continued*
 resiliency of, 171
 rings, 230
 sales, 169
 scrap, 168
 seeds, oil from, 121
 seeds, winning of, 6
 sheet, 225
 shoes, 230
 smoking of, 165
 soils, analyses of, 94
 solution, 230
 solvents, Weber on, 213
 synthetic, 214 *et seq.*
 vines planted in Congo, 209
 washing of, 159

Sales, rubber, 169
Schidrowitz, 36
 on intercrops, 114
 on reclaimed rubber, 235
Science, applied to pests and diseases, 177
Scott, Dr D. H., 47
"Scrap, Ceara," 25
 rubber, 168
Seasonal variation in latex, 71
Seed bearers, 102
 selection, 101
 selection by progeny, 103
Seedling, laticiferous system in, 46, 47
Seedlings, in baskets, 100, 110
Seringa, 4
Seringal or estate, 22
Seringueiro or collector, 4, 20
Shade belts, 119
Sheet rubber, preparation of, 225
Siam, 35
Singapore, 7, 8
Smoked sheet, 164
Smoking, 165
 of wild rubber, 20
South India, labour in, 122
Spaniards, early employment of rubber by, 3
Species of rubber, American, 17 *et seq.*
 African, 30 *et seq.*
 Asiatic, 34 *et seq.*

Sphaerostilbe repens, 184, 187
Spiral system of tapping, 49
Spur-shaped pricker, 132
Stagbrook Rubber Co., crop of, 73
Stem, diseases of, 187
Stone cells, 44, 49
Stumping, 99
Sudan, 33
Sulphur, in rubber manufacture, 224
Sumatra, 35, 36
Synthetic rubber, 214 *et seq.*
 discovery by Bouchardet, 215
Syringe tree, 4
Syringes, 4

Tackiness, 175
Tapping, age for, 128
 average yields (grammes), 77
 basal system, 139
 difficulties, 149
 effects of, 90
 excision methods of, 136
 experiments, 56
 Fitting on, 48
 herring-bone system, 133, 139
 ideal rate of, 80
 incision methods of, 130
 increase of yield on, 61
 intervals, 62, 75, 149
 methods of, 57
 moderate, 90
 paring process, 143
 precautions, 149
 rules for, 92
 severe, 90
 spiral system of, 49
 systems of paring, 137
 time occupied in, 76
 tools, 144
Tappings, results on intervals of, 62
Tea, as intercrop, 114
Teats, 230
Tephrosia, 113
Termes gestroi, 180
Thinning out, 121
Tilden obtains synthetic rubber, 214
Tisdall, W. N., estimate by, 127
 on stumping, 109
Tobacco pouches, 228

Tools, for tapping, 144
    used in Ceylon, 126
    used in Malaya, 126
Torquemada, Juan de, 3
Transport in Brazil, 23
    in the East, 97
    of latex, 156
Trimen, Dr, experiments in tapping
    Hevea, 8
    pricking by, 131

Ule rubber, 26
Ulequahuitl (*Castilloa elastica*), 3

Vacuum driers, 163, 165
    in manufacture of rubber, 223
"Vapourite," 181
Vegetative organs, 38
Vulcanisation, 2, 216 *et seq.*
    cold, 218, 233
    Goodyear's dry process, 231
    hot, 217
    in Amazon districts, 230
    Hancock's wet process, 231
    Parkes' cold process, 231, 233
    time required for, 232
    Weber on, 217
Vulcanite, 236

Wages, in Ceylon, 125
    in Malaya, 125
Washing machines, 160
Water supply, effect on latex, 88
Weber, on rubber solvents, 213
    on vulcanisation, 2, 217
Weeding, 111
West Africa, rubber planting in, 11
West Indies, 5, 7, 8
Wickham, H. A., 5
    brings *Hevea* seeds to Kew, 6
    smoking machine of, 166

Wild rubber, African, 30 *et seq.*
    from Africa and Asia, 4
    from Congo State, 5
    replanting in Brazil, 209
    smoking of, 20
    tapping of, 20
    trade in, 4
Willis, Dr J. C., 56
    experiments by, 60
Willughbeia, 35, 36
Wind, affects *Hevea*, 179
    belts, 119
Wound response, 59
    experiments in Ceylon, 60
    in *Castilloa*, 59
    in *Funtumia*, 59
    in *Hevea*, 59
    in *Manihot*, 59
    increased yield attributable to,
        61
Wright, Herbert, estimate by, 12
    on pricking, 130

Yield, at different levels, 80, 81,
    148
    duration of, 64
    effect of rainfall on, 73
    from Henaratgoda tree, 66, 67
    from tapping at frequent in-
        tervals, 78
    general considerations affecting,
        87
    in relation to bark, 66
    in relation to volume of bark,
        66
    in grammes, 77
    per acre, of dry rubber, 64
    per tapping, variation in, 72
    plantation, 151
    season of highest, 72
    variation in, 74

Printed in the United States
By Bookmasters